INTERNATIONAL SERIES OF MONOGRAPHS
ON PURE AND APPLIED BIOLOGY

Division: **ZOOLOGY**

GENERAL EDITOR: G. A. KERKUT

VOLUME 13

ELECTRON-MICROSCOPIC STRUCTURE OF PROTOZOA

OTHER TITLES IN THE ZOOLOGY DIVISION

General Editor: G. A. KERKUT

OTHER DIVISIONS IN THE SERIES ON
PURE AND APPLIED BIOLOGY

BIOCHEMISTRY

BOTANY

MODERN TRENDS
IN PHYSIOLOGICAL SCIENCES

PLANT PHYSIOLOGY

ELECTRON-MICROSCOPIC STRUCTURE OF PROTOZOA

by

Dr. D. R. PITELKA

Cancer Research Genetics Laboratory
University of California, U.S.A.

PERGAMON PRESS

OXFORD · LONDON · NEW YORK · PARIS

1963

PERGAMON PRESS LTD.
Headington Hill Hall, Oxford
4 & 5 Fitzroy Square, London, W.1

PERGAMON PRESS INC.
122 East 55th Street, New York 22, N.Y.

GAUTHIER-VILLARS
55 Quai des Grands-Augustins, Paris 6

PERGAMON PRESS G.m.b.H.
Kaiserstrasse 75, Frankfurt-am-Main

Distributed in the Western Hemisphere by
THE MACMILLAN COMPANY · NEW YORK
pursuant to a special arrangement with
PERGAMON PRESS LIMITED

563
Pit

Library of Congress Card No. 62-19274

Set in Garamond 11 on 12-pt. and printed in Great Britain by
THE BAY TREE PRESS, STEVENAGE, HERTS

This book is affectionately dedicated to

the memory of Harold Kirby

CONTENTS

ILLUSTRATIONS

PREFACE

ATTEMPTING an analytical review of a rapidly developing field is rather like eating spaghetti with a spoon. In order to manage at all, one has to cut some strands off short, while others elude one altogether. This is so even though the total serving is not unreasonably large; in 1961 the literature on electron-microscopic structure of protozoa is still within the capacity of one reviewer. The organisms known as protozoa include such a diversity of types, however, that no two protozoologists (who are also diverse organisms) approach them in quite the same way.

In coping with this slippery prospect, I have emphasized details that seem to me to be provocative or significant, and understated or ignored others that to another critic may appear startling. Inevitably, many of the questions raised in the following pages are being answered through investigations now in progress; certainly some of the answers will be apparent before the book reaches its readers, and new problems not yet visualized will be posed.

But the yearning for an occasional summing-up is no less acute because the field is in flux. Whether we approach the problem as protozoologists (of whatever ilk), as evolutionists, or as cell biologists, we should like to know whether results from ultra-structure research suggest that the protozoan grade of organization embodies any peculiar morphological features. Before the mush-rooming quantity of data becomes too unmanageable, it is time to stop and take a look.

Although I have aimed at a complete coverage of the immediately pertinent literature, there will be omissions, for which I express my regret. Papers published in specialized journals of parasitology, medicine, or botany are most likely to have escaped my attention. Only a sampling has been included of taxonomic papers drawing on electron microscopy for details of superficial structure as criteria for species distinctions. Where information presented in a preliminary report or abstract is repeated or amplified in a more complete later publication, the earlier one is not cited unless priority is importantly involved. With the exception of a few papers that were made available to me in manuscript, the coverage of literature terminated with works published in June, 1961.

The original electron microscopy reported here was supported in part by grants C–5388 and C–4108 from the National Cancer Institute, Public Health Service. I am indebted to many colleagues for indispensable co-operation in the preparation of this book. For their generosity in supplying me with original micrographs — many of them from unpublished work — and for permission to reproduce diagrams, I am deeply grateful to Everett Anderson, H. W. Beams, L. H. Bretschneider, Joan E. Cook, Ruth V. Dippell, Charles F. Ehret, E. Fauré-Fremiet, I. R. Gibbons, P. P. Grassé, J. Ludvik, Charles B. Metz, Cécile Noirot, Béla Párducz, Pierre de Puytorac, J. T. Randall, L. E. Roth, C. Rouiller, Maria A. Rudzinska, Ruth Sager, L. Schneider, and K. E. Wohlfarth-Bottermann. I owe thanks for expert photographic assistance to Jerome Baron; for patient persistence in typing and editing to Eleanor Burns; and for help in ways too numerous to mention to Carolyn G. Smoller, K. K. Sekhri, Joanne Rush, and particularly to my husband, Frank A. Pitelka. Finally, it is a pleasure to express my sincere gratitude for their critical reading of all or parts of the manuscript to William Balamuth, Ellsworth C. Dougherty, and Daniel Mazia.

CHAPTER 1

INTRODUCTION

" . . . The history of the living world can be summarised as the elaboration of ever more perfect eyes within a cosmos in which there is always something more to be seen."

P. Teilhard de Chardin, *The Phenomenon of Man,* 1959

"(J'ai) . . . beaucoup d'observations au microscope électronique, mais il manque toujours celles qui doivent éclairer toutes les autres."

E. Fauré-Fremiet, letter to the author, 1960

TODAY's biologist has at his disposal eyes more perfect in their capacity to see than anything his predecessors could imagine. Yet as always, organisms have to be tricked into exposing their secrets to his view, and the arts of this trickery inevitably lag behind the development of viewing tools. Presented rather abruptly only two decades ago with the marvelous bonanza that is the electron microscope, the biologist has had to undergo the painful process, repeated anew with each new kind of material, of learning how to trap his subjects in a state fit to be seen. And once he has accomplished this with reasonable success, he finds himself in a subcellular world that has suddenly become enormous, wherein he needs something more than just seeing to learn his way around.

So it is that, for all the grandeur of the prospect, every electron microscopist finds himself sooner or later — if not both — in the position wryly described by Professor Fauré-Fremiet. We may yield with thorough and joyous enthusiasm to the pleasures of looking with our new eyes. But our comprehension of what we see depends on a constant awareness of the vastness of the area that has not yet been seen.

For some of the very same reasons that have caused most biologists of recent generations to leave them in limbo, the protozoa offer material of unsurpassed excellence to the electron microscopist. Where their small size once drove every investigator

1

against the uncomfortable boundaries of light-microscope resolution, this same quality now has the advantage of encompassing a multitude of structures and functions within a single package that the electron microscope can open. Because each cell is also a relatively independent organism, groups of individuals can be studied in parallel as fully and naturally functioning units and as assemblages of structural parts.

From ancestors whose identity is obscured by time, the earliest protozoa inherited the ability to segregate many vital activities within membrane-bounded intracellular compartments, and they were ingenious experimenters with this equipment. They explored countless possibilities and achieved enough successful combinations to give rise to the metaphytan and metazoan lines as well as to the myriad forms that have survived admirably without ever finding it necessary to conscript battalions of interdependent units. During the course of this experimentation they hit upon most of the special functions that are differentially assigned to less versatile cells, tissues, and organs in the metazoa and metaphytes. In addition to the every-day miracles that each living cell must perform, protozoa perfected such remarkable skills as contraction, skeleton formation, co-ordinated locomotion, sexual reproduction, photosynthesis, symbiosis both benign and pathogenic, and the morphogenesis of complex form and pattern.

As innovators, specialists, models, and extremists, the protozoa have been exploited for only a tiny fraction of the information that is ours for the seeking. While we are beginning to feel rather comfortably at home within the minute dimensions of some of the more domesticated vertebrate cells, there is not yet a single protozoan type whose ultra-structure and ultra-function have been so thoroughly mapped. Too many of the clarifying micrographs are still wanting, even for the most familiar species, and the combined application of electron microscopy with other, analytic techniques has rarely been attempted. The generalizations that we can begin to make about protozoa are based on comparisons with "higher" organisms, and it ought to be the other way around.

The fraction of information on protozoan ultrastructure that we have obtained, and the tentative generalizations emerging from it, are the subject of this book. Since representatives of most of the major protozoan groups have been studied, it is possible to outline a patchwork comparative anatomy. This will continue to be patchwork for a long time, but a review now should point out the gaps most conspicuously in need of filling. Moreover, by assembling some of the remarkable facts of protozoan structure, such a review may demonstrate to the cell biologist that the educational value of the protozoa, as proclaimed above, is not exaggerated. (For similar arguments from evangelists of biochemical protozoology, see for example Hutner, 1961; Hutner and Provasoli, 1951.)

In what follows, the approach will be fundamentally evolutionary, based on the proposition that all eucellular* life had a common origin in a progressive moneran* stock. In agreement with the majority of modern protozoologists (see Hyman, 1940; Grassé, 1952) it is assumed that the phytoflagellates are the oldest protozoa. Recently there has been a spate of thoughtful and often elegant discussions of the origin of life and the mechanisms of evolution (e.g., see Needham, 1959; Gaffron, 1960). Evolutionary histories of many groups have been considered, but relatively few novel speculations on the phylogeny of eucellular protists* have been offered. The one explicit new theory is that of Dougherty (1955) and Dougherty and Allen (1960), founded largely on evidence from biochemistry. They suggest, in brief, that phytoflagellates arose from primitive red algae, which in turn derived from blue-green algae (a similar idea is implied, but not discussed, by Copeland, 1956). Evidence from electron-microscope studies that bears on, and tends to support, this theory will be discussed in Chapter 4.

Arguments in favor of a common ancestry of eucells rest largely on similarities in biochemistry and in the ultrastructure of common organelles. An opposing point of view, based on the same evidence, is held by Kerkut (1961) and Grimstone (1959b, 1961).

---

* The present usage of these terms, not original here, should be explained. *Monera* (= bacteria and blue-green algae) and *Protista* (= Monera plus all algae, fungi and protozoa) are used as defined most recently by Dougherty and Allen (1958). *Eucell* (= cell provided with a membrane-limited nucleus or *eukaryon* [Dougherty, 1957a]); in other words, cells of all organisms above the Monera) is used as proposed by Picken (1960).

These authors argue that such similarities may be the result of inescapable convergence. Recognizing the inventiveness of living matter and the fixity of physical laws, they suggest that totally unrelated stocks of primitive organisms could have hit upon the same solutions — the only possible solutions — to the problems of survival and reproduction.

The decision whether a ubiquitous solution is the only possible one or instead one that, at the time of its invention, gave its possessors a literally overwhelming selective advantage will require a far deeper understanding of biological evidence than we now possess. At present the same deductive arguments can be piled up on both sides of the fence. Chloroplasts, mitochondria, Golgi elements, nuclear membranes, flagella, centrioles — all show an impressive universality of structure. They are morphological expressions of ubiquitous solutions to problems that we can define only in general terms. Are they the results of a common genetic heritage or of repetitive, independent, evolutionary events? The study of protistan comparative morphology can at the very least contribute circumstantial evidence toward an answer.

To consider all the implications of protozoan ultrastructure would require a review of the whole field of cell biology, as well as of all protozoology. The author disclaims any such ambition and recommends as sources of relevant information competent reviews and treatises in cell biology and ultrastructure such as those by Brachet (1957), Picken (1960), and numerous authors under the editorship of Palay (1958) and of Brachet and Mirsky (1959 et seq.), and in protozoology those by Grassé and others (Grassé, 1952, 1953), Hyman (1940), Hall (1953), Kudo (4th ed., 1954), Smith (1955), Grell (1956), and Schussnig (1960). It hardly needs emphasizing that the work to be reviewed here would be meaningless without the solid edifice of light-microscope studies in protozoology to support it.

Argument over the limits of the unnatural "Phylum Protozoa" will be evaded by accepting as protozoa all those organisms so recognized in Volume I of the *Traité de Zoologie* (Grassé, 1952, 1953) and by following the classification proposed therein (except that uniform endings for taxon names, as specified by Hall, 1953, will be imposed on Grassé's system; see Appendix). The author recognizes (as does Grassé) that this delimitation arbitrarily cuts across certain natural protistan relationships (such as that between the more advanced multicellular algae and their motile, unicellular, "protozoan" relatives, and, with even less logic, that between certain groups of unicellular forms — diatoms, desmids, and other non-motile chlorophytes — and related forms here treated as "protozoa"). The justification for this is the purely pragmatic one of having had to establish limits in order to get the job of writing this book done, combined with the author's own interest in animal-like forms and their ancestors. Of course it often will be necessary to draw upon comparative data about non-protozoan protists, as well as metaphytes and metazoa, but such data will be cited only to clarify or amplify available information on protozoa.

Techniques and Interpretations in Electron Microscopy

It is not necessary here to reiterate the potentialities and problems of electron microscopy. This has been done repeatedly. Recently, Schmitt (1960) has presented a judicious evaluation of the position of the electron microscope in the spectrum of powerful modern biophysical and biochemical tools and has reviewed some of the methods by which data accruing from these may be cross-checked. As familiarity with the literature in electron microscopy increases, the reader who asks the question "How far can I trust what I see — or what the author says I see?" needs a set of criteria neither different in principle nor more mysterious than those he would apply to evidence from any other source. These are essentially the same criteria applicable to light microscopy, but magnified. The most important single qualification is that the electron micrograph illustrates a structure artificially fixed in a moment of time. It can yield information on cell dynamics *only* if carefully controlled ancillary techniques are used.

Almost all of the information to be discussed is drawn from protozoan cells fixed in buffered solutions of osmium tetroxide, dehydrated, embedded in plastic or resin, and thin-sectioned. These techniques were first employed little more than ten years ago; since then they have improved so rapidly that, in general, more recent reports in this short interval can be assumed to be more accurate than earlier ones (although of course improvement has not been simultaneous everywhere). The requirement for perfection in preparative techniques varies with the questions that are being asked. If a strongly bonded, insoluble protoplasmic structure is under investigation, it may survive the ordeal of preparation and yield valuable information while a neighboring, more delicate constituent is distorted or destroyed. Fibrous organelles are likely to be particularly sturdy.

Minor variations in temperature, osmotic pressure, pH, and timing of fixation may affect different kinds of cells, or of structures within cells, quite differently; hence the desirability of exploring alternative fixing methods for each new material. The composition of the embedding medium is another significant variable; the most commonly and easily used plastic, methacrylate, occasionally damages cells during the hardening process. Other media (Vestopal, epoxy resins) are being substituted with increasing success.

The recent emphasis on thin sectioning has tended to obscure the valuable results that may be achieved by other methods. Particularly where orderly patterns are spread out over a cell, thin sectioning exposes too minute an area to permit detection of the overall pattern. In these cases, surface replicas may be informative; methods of fragmentation or progressive dissolution often yield isolated fragments or pellicle strips that are essential for the accurate interpretation of sections.

Non-specific staining of cells before or after sectioning is often helpful in increasing contrast in the final image. Heavy-metal solutions such as phosphotungstic acid, lead hydroxide, or uranyl acetate are among those most commonly used. The development of specific procedures for electron-microscope cytochemistry is well under way; these have rarely been applied as yet to the protozoa, since they always require a thorough preliminary familiarity with normal morphology.

Measurements made on electron micrographs are of inconsistent value. Examples will appear in which one author reports the diameter of a given sort of fibril, say, as 8 mμ, while another measures what should be precisely the same structure as 15 mμ. The most commonly used methods of calibration of the microscope are recognized to be approximate, and final measurements are subject to considerable individual error. At this stage in protozoology, these variations are not usually of great importance. Relative sizes may be estimated, and as long as we are dealing with structures of unknown molecular composition, subjected to unknown alterations during preparation, the significance of differences in minute dimensions is remote. Only for a very few types of materials (striated muscle, collagen, myelin) are we in a position to care deeply whether a given interval is measured at 5 mμ or at twice that. To avoid any implication that figures are accurate to the last Ångström unit, all measurements will be given here in microns and millimicrons.

CHAPTER 2

PROTOZOA AS CELLS

WITH major emphasis being placed today upon "the cell" as a structural and functional entity, it is wise to bear in mind Hyman's (1940, 1959) insistent warning that a protozoon is better to be compared with a whole metazoan organism than with any of its component cells. As significant as this concept is, however, there are contexts in which it becomes unmanageable, and the chief of these is cytology (a point well made by Grimstone, 1961, in concluding an admirable short review of protozoan ultrastructure). The electron microscope has demonstrated that the fine structure of protozoa is directly and inescapably comparable with that of cells of multicellular organisms. Where the greater versatility and self-sufficiency of protozoa is accompanied by a higher degree of structural elaboration, this will become apparent — even startlingly so — in a comparison of such protozoa with other cells. In other instances, what is startling is the slight morphological differentiation of protozoan organisms. The morphologist has to start out by admitting that protozoa are, at the least, cells.

In this chapter the morphological properties that are common to many or most protozoa — and to many or most other kinds of cells — are considered. Particular emphasis is given to organelles such as contractile and food vacuoles that are especially, though not uniquely, protozoan. Ameboid locomotion — an unquestionably fundamental phenomenon — will, however, be discussed in the following chapter, for the practical reason that it is not possible to discuss it without describing the structure of the entire ameba.

Membranes and the Cytoplasmic Matrix

Not surprisingly, the surfaces of protozoan cells often are peculiarly and elaborately specialized on an ultramicroscopic scale as well as at the familiar level of light microscopy. Fibers, folds, blisters, and scales embellish different species in characteristic patterns. Often these differentiations involve the cell membrane itself or are intimately layered below it, so that long-standing questions as to whether pellicles or cuticles are living structures can for some species be answered in the affirmative. Descriptions of these widely varying characters will be given in the taxonomic sections to follow.

The one generalization we can make is that most if not all protozoa have at the actual surface of the protoplasm a limiting membrane that shows the same structure as has been reported for other cell types. Such a plasma membrane is best described as a three-ply sheet; it appears in cross-section as two dark lines, each about 2 mμ thick, separated by a light zone. The total thickness of the membrane is about 7 to 8 mμ (Bennett, 1956; Sjöstrand, 1959; Robertson, 1960). Comparison of this structure with the molecular arrangement proposed on other than morphologic grounds has led to the assumption that the three-ply construction represents a bimolecular leaflet of mixed lipid molecules whose non-polar ends face each other and whose polar surfaces are covered by thin layers of non-lipid material. The varying nature of this material, probably protein in most cases, is accountable for the highly individual properties of membranes and for their chemical asymmetry, as well as for some of the variations in apparent thickness or density of membranes seen in electron micrographs.

If the conclusions drawn by avid students of membrane structure are correct (see Robertson, 1960), *all* protoplasmic membranes, not just that at the cell surface, share this fundamental molecular architecture; the three-ply sheet is considered to be the unit membrane.

The importance of membranes within as well as around cells is no news to cell physiologists, but their abundance on an ultramicroscopic scale and their elaborate involvement in many familiar organelles have been among the major discoveries of electron microscopy (see Text-fig. 1). It is becoming clear that membranes provide the framework upon which orderly sequences of reactive compounds are applied. Complex chemical processes can thus be handled by an assembly line of fixed enzymes; the concept of metabolic "pathways" may be more literal than had been supposed.

More or less permanent, heavy-duty organelles, such as mitochondria, nerve myelin, chloroplasts, and the outer segments of vertebrate photoreceptors, are fairly easily comprehended in these terms. For the more evanescent membranous structures of the cell, the picture is not so clear. Detection of the three-ply form requires a high degree of fidelity in preservation and micrography, as well as a lucky angle of sectioning. Viewing the static picture of fixed structures provided by the electron microscope, one can too easily forget the visible activity of the living cell. Here the protozoologist's necessary preoccupation with his organisms as living entities may serve him in good stead. For example, the almost explosive formation of pseudopodia by many ameboid cells indicates a virtually instantaneous production or extreme stretching of surface membranes. Extensive but, up to a point, reversible vacuolization of the cytoplasm in many protozoa when subjected to unfavorable conditions on a microscope slide is a commonly observed phenomenon; the cytoplasm-vacuole interface here appears as a membrane in electron micrographs, and certainly represents an area vastly increased over what was present before. And the incessant streaming of organelles within the protoplasm of active cells must preclude the existence in these regions of an architecturally fixed network of membrane-bound spaces. In other words, many of the membranous structures we see in electron micrographs must be transient in form and position if not in construction. One cannot imagine how a cell, and particularly a free-living one, could survive if the production, renewal, and elimination of membranes were not to be accomplished with great facility and with materials readily at hand.

Much of this chapter is devoted to a consideration of membranous organelles as they appear in protozoa. First, however, we must consider the ground substance (which must contain, among other things, membrane precursors). The cytoplasmic matrix, following the rigors of preparation for electron microscopy, appears to consist of very fine filaments and/or granules dispersed in a low-density continuum; these probably include both structural elements and precipitates of protoplasmic components left more or less *in situ* as the protoplasmic water and low-molecular-weight solutes were withdrawn during dehydration of the specimen. In a significant and meticulous recent study on amebae, Wohlfarth-Bottermann (1961) has applied a series of different fixing agents to cells grown under identical conditions. These included the conventional osmium tetroxide, osmium tetroxide plus potassium dichromate, potassium permanganate, formalin alone, and formalin preceded or followed by osmium tetroxide. He found that the same recognizable structure, and variation in structure, can be demonstrated following any one of these fixatives. The appearance of the hyaline cytoplasm varies, within any single cell, from nearly homogeneous through a condition marked by the random dispersion of minute (8 to 9 mμ) globular particles to a compact net-like arrangement resulting apparently from the linear aggregation of globular particles, which now appear larger and denser. The last-named aspect he believes represents a state of gelation or higher viscosity of the cytoplasm.

One class of cytoplasmic granules is of particular significance. These are dense, irregularly angular in shape, and 10 to 15 mμ in diameter (Fig. 1, Pl. I); often they are disposed on membrane surfaces. The microsome fraction of centrifuged cell homogenates, which has distinctive biochemical properties, is composed largely of particle-studded membrane fragments and free particles. Palade and Siekevitz (1956a, b) first succeeded in identifying these particles, from rat liver cells and from pancreatic exocrine cells of the guinea pig, as ribonucleoprotein (RNP), and it is now generally agreed that they represent the principal basophilic component of cytoplasm. It cannot be assumed, of course, that all granules of this description are RNP, and not all ribonucleic acid (RNA) is confined to the particles, but correspondence between particle frequency and cytoplasmic basophilia is a general rule. Whether the particles represent a precipitate or are present as such in life is not certain. Dense, angular particles within the size range 10 to 15 mμ or thereabout will be referred to here descriptively as Palade particles, without intending to imply that their composition is known.

Scattered through this matrix are membrane-enclosed compartments ranging from tiny vesicles and canaliculi through extensive flattened sacs to large vacuoles, in a continuous spectrum of sizes

PLATE I

FIG. 1. Section of cytoplasm of a small zooflagellate, *Bodo saltans*, showing abundant Palade particles, granular membranes, and several sections through a single mitochondrion. × 45,000.

FIG. 2. Section through one mitochondrion and parts of two others in the ciliate *Colpidium campylum*, illustrating microtubular configuration of internal membranes and, in mitochondrion at bottom, a finely filamentous inclusion. Double membranes surrounding mitochondria are distinct in some areas; elsewhere they are cut obliquely. × 45,000.

FIG. 3. Section through several Golgi bodies composing part of parabasal apparatus of the complex zooflagellate, *Joenia annectans*. At proximal end of each pile (arrows) is parabasal fibril cut in cross-section. Cluster of bodies occupying upper half of figure represents a probable method of Golgi reproduction. Four discrete piles are evident proximally, each with its own parabasal fibril. Distally, several sacs are continuous through three or all four piles, suggesting that one original body is being segmented into four. × 31,000. From Grassé, 1957c.

FIG. 4. Section through mitochondrion of *Euglena gracilis*, showing radially arranged cristae. × 50,000. From Gibbs, 1960.

FIG. 5. Section of *Bodo saltans*, showing bulbous kinetoplast at center, continuous at left and right with mitochondrion. × 22,500.
From Pitelka, 1961b.

PLATE I

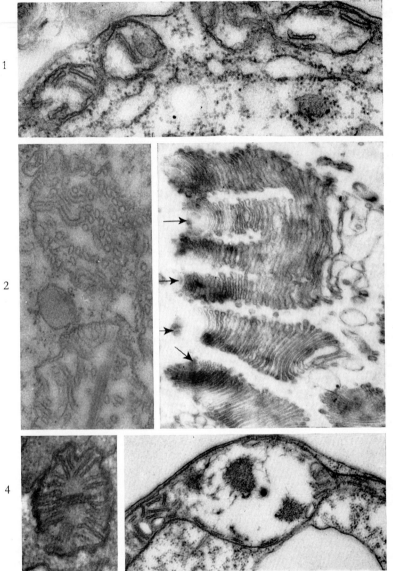

and shapes. Certain consistent configurations of membranes constitute identifiable organelles that will be described presently. It has become common practice to refer to any other small membranous elements as "endoplasmic reticulum" (Text-fig. 1),

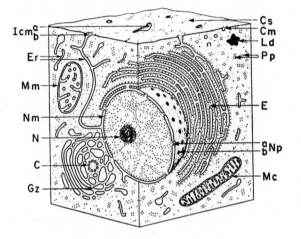

TEXT-FIGURE 1. Diagrammatic reconstruction of part of idealized cell containing organelles characteristic of many cell types. From Wohlfarth-Bottermann, 1960a: Icm, Invagination of cell membrane, a. surface view, b. section; Er, Endoplasmic reticulum; Mm, Mitochondrion (microtubule type); Nm, Nuclear membrane; N, Nucleolus; C, Centriole; Gz, Golgi zone; Cs, Cell surface; Cm, Cell membrane; Ld, Lipid droplet; Pp, Paladc particles, E, Ergastoplasm; Np, Nuclear pore, a. section, b. surface view; Mc, Mitochondrion (crista type).

whether or not they are exclusively endoplasmic or demonstrably reticular. This term was introduced by Porter and Kallman (1952) to describe a system of tubes and sacs ramifying through the cytoplasm of thin whole mounts of tissue culture cells. Because of this observation, and because various membranous structures are frequently seen in sections to be interconnected (very extensively so in some higher plant cells), it has been suggested that the entire protoplasm is segrated into a continuous phase and a membrane-enclosed phase, and that the membrane-limited spaces are at least potentially all in communication with each other and with the outside environment of the cell.

Cytoplasmic membranes always surround and enclose something, even though the enclosed space may become nearly eclipsed by flattening. They are, as would be expected, phase boundaries and their formation and function must be intimately related to the existence of the two phases. At least some of them are exceedingly labile structures, yet their molecular framework cannot be haphazard and (if the unit membrane concept is valid) is always basically the same. The cytoplasmic matrix must contain in considerable abundance molecules capable of being rapidly incorporated into existing membranes or aligned to form new ones; it would seem economically sound if already formed membrane skeletons could, where available, serve as templates in this process. The specific nature of the membrane surface facing the cytoplasmic continuum should be roughly the same in adjacent regions of the cytoplasm; the specific nature of the surface facing the enclosed phase should vary with the source and composition of that phase. Temporary or permanent fusion of neighboring membranous compartments could under these circumstances be expected at least accidentally and might subserve important functions of transport. Where very precisely predetermined enzyme assemblages are required to carry out specific localized reactions, membrane packages of a high degree of order, such as mitochondria, chloroplasts, etc., are formed, and these are generally more stable.

What all this means is that for most cells, and particularly for most protozoa, the significance of miscellaneous cytoplasmic membranes is not known. When we use the term "endoplasmic reticulum" for such membranes we use it descriptively, recognizing that the structures may be evanescent, and without implying any certitude about their source, fate, or function.

In a majority of cells, including protozoan, at least some of the cytoplasmic membranes are studded on their outer (facing the cytoplasmic matrix) surfaces with Palade particles (Fig. 1, Pl. I). The outer surface of the nuclear envelope is typically granular and often is continuous with granular endoplasmic reticulum. In several types of metazoan cells that are actively engaged in elaborating protein secretions for export, the granular reticulum takes the form of impressive arrays of concentric double lamellae forming flat sacs or cisternae. Such arrays may be called ergastoplasm (thus the terms ergastoplasm and endoplasmic reticulum overlap broadly; for a history of the concepts and terminology see Hagenau, 1958; and Porter, 1961). Ergastoplasm is extensively developed, for example, in pancreatic acinar cells (Palade, 1959) and in lactating mammary epithelium (Wellings, DeOme, and Pitelka, 1960); it has also been found in several kinds of invertebrate cells (*e.g.* the developing cnidoblasts of *Hydra*, Slautterback and Fawcett, 1959), but where it occurs in protozoa it usually is limited to scattered sacs or a few parallel cisternae, and a relationship to protein synthesis has not been demonstrated.

The Golgi Apparatus

Electron-microscope studies have put an end to a long-continuing controversy over the reality of the Golgi apparatus by proving that it is a well-defined membranous entity present in most if not all eucells. It consists (Fig. 3, Pl. I) of small to large piles of thin, flat sacs, never bearing Palade particles — but occasionally in continuity with granular cytoplasmic membranes. Minute vesicles clustered at the edges of the sacs appear to be pinched off from their rims. Larger vacuoles represent inflated single sacs. The dictyosomes of animal spermatocytes, the dictyosomes and true parabasal bodies of flagellates (but not all structures that have been called parabasals), and the Golgi zones of metazoa all fit this description, and similar structures occur in plant cells and in many protozoa not previously supposed to contain them.

Grassé and his colleagues must be credited for much of the work of bringing the identification of Golgi elements in protozoa into line with the picture seen in metazoan cells. Following earlier light-microscope studies (reviewed by Grassé and Hollande, 1941), they sought and found Golgi structures in electron micrographs of gregarines (Théodoridès, 1959), and of ciliates (Noirot-Timothée, 1957) as well as of a variety of flagellates (Grassé, 1956a, 1957a, 1957b, 1957c; Grassé and Carasso, 1957). Grassé (1957c) further has characterized the Golgi apparatus as a distinctly polarized structure. This is particularly clear in some zoo-flagellates, where a striated fiber originating at the centriole terminates on the parabasal body. The Golgi sacs at this proximal side of the pile are thin and sharply defined (Fig. 3, Pl. I). At the opposite side the sacs are less orderly and frequently inflated. Grasssé suggests that these sacs are in the process of breaking away as swollen vesicles, carrying their contents off into the cytoplasm, while their replacement is assured by constant new formation of sacs at the proximal end. The proximal sacs would represent the chromophilic component of the Golgi complex as seen in cells stained for light microscopy; the detached, swollen vesicles at the distal side compose the chromophobic zone.

An illuminating study of protozoan Golgi bodies and their possible relation to other membranes is that by Grimstone (1959a, 1959c) on the complex zooflagellate *Trichonympha*. In this genus the outer surface of the nuclear membrane is, as usual, studded with granules, and a corona of granular membranes under normal conditions surrounds the nucleus. The parabasal bodies are clustered about the nucleus, polarized as described by Grassé, and with their proximal faces always directed toward the nuclear surface. By withholding food from their termite hosts, Grimstone was able to observe the effects of starvation on the wood-feeding *Trichonympha*. Starvation resulted within two days in a complete disappearance of granular membranes; at the same time the parabasal bodies came to consist of segmented, inflated sacs only. After three days, some parabasal filaments were seen associated with only a few swollen sacs. Twenty-four hours after refeeding, granular membranes were present but sparse, and parabasal bodies consisted of exceptionally large piles of flat, oriented sacs. Following another day with food, granular membranes again were abundant and parabasals contained a normal complement of flat and inflated sacs. Grimstone suggested that the granular reticulum was contributing membranes to the parabasal bodies. His cytochemical studies showed that the parabasal apparatus contained acid phosphatase and a polysaccharide which also appeared in some abundance near the cell membrane. He postulated that the Golgi zone may function to produce a poly-saccharide that could have unusual structural importance in these flagellates with low protein resources.

Biochemical studies by Kuff and Dalton (1959) on Golgi membranes isolated from rat epididymis indicated a concentration there of lipid phosphorus and acid phosphatase, but no unique biochemical properties were found. In general, the assumption that the Golgi apparatus functions in secretory activities is borne out by its appearance in many vertebrate cells where, for example, granules of zymogen (Palade, 1959) or of milk protein (Wellings and DeOme, 1961) accumulate within Golgi membranes to form aggregates that may then move elsewhere in the cell, still enclosed in Golgi-derived vacuoles. In most protozoa — and most other cells where secretion products are less conspicuous or less intensively produced — good clues to the function of the Golgi complex are lacking.

Mitochondria

Mitochondria are another kind of membranous organelle that has a highly characteristic and remarkably uniform structure in all eucells. Whatever their size or shape, mitochondria are limited by two unit membranes, separated by a gap of about 10 mμ. Internally there are additional membranes arranged usually in the shape either of flat shelves called cristae or of convoluted microtubules. These appear to be continuous with the inner one of the two limiting membranes, as though they were formed as internal extensions of it. Variations in the abundance and in the orientation of the cristae or microtubules are common; rarely, very orderly patterns are found. The matrix surrounding the internal membranes varies in density and has little visible structure, but inclusions often appear in the form of small dense bodies or bundles of fine filaments.

Mitochondria in most cells of metazoa and of land plants have internal cristae. Microtubular mitochondria (Fig. 2, Pl. I) are characteristic of the majority of protozoan types studied to date, but also occur in an odd assortment of other cells (Sedar and Rudzinska, 1956; Rouiller, 1960; Novikoff, 1961). Among the flagellates, both types, and perhaps some intermediates, are seen. Chrysomonad flagellate mitochondria are microtubular; most green algae and euglenoids (Fig. 4, Pl. I) examined have cristae. In a relatively simple zooflagellate, *Bodo,* a mitochondrion of rather curious structure occurs (Pitelka, 1961b). Internal membranes have the form of randomly oriented flat discs (Figs. 1 and 5, Pl. I); only rarely do they appear to be in continuity with the inner limiting membrane. In *Trichonympha* (Pitelka and Schooley, 1958) and in *Opalina* (Noirot-Timothée, 1959), both intestinal symbionts, bodies with a much finer tubular internal structure than usual are tentatively identified as mitochondria.

Pappas and Brandt (1959) found in the giant ameba, *Pelomyxa carolinensis,* at the time of nuclear division, some mitochondria with a conspicuous pattern of organization. Its main theme consisted of microtubules with regular zigzag profiles. Where parallel tubules were in phase, a strikingly constant space, about 25 mμ wide, separated them; the space within the tubules was wider and more variable, 40 to 70 mμ. Elsewhere, adjacent out-of-phase zigzags apparently fused at their peaks to produce regularly fenestrated plates, and three-dimensional interdigitation of neighboring zigs and zags produced complicated but orderly geometric patterns. Somewhat similar configurations have been illustrated in the mitochondria of developing spermatids (André, 1959).

No explanation for the occurrence of these remarkable patterns is available. Nor is any consistent functional significance ascribable to the difference between cristal and microtubular form in the internal membranes. Osmotic properties of microtubular mitochondria isolated from *Acanthamoeba* (Klein and Neff, 1960) proved similar to those of mammalian mitochondria. Vickerman (1960) found that *Acanthamoeba* mitochondria contained dense inclusions within the microtubule lumina, and that these were significantly more numerous in cysts than in active cells. Stewart and Stewart (1961) similarly have reported abundant dense inclusions in the mitochondria of the slime mold *Physarum* in the dormant, sclerotized phase. But the significance of the inclusions is unknown.

Mitochondria are known to be the principal site of oxidative phosphorylations in aerobic cells (Lehninger, 1959; Green and Hatefi, 1961). Many enzymes of the respiratory chain are bound on or within the mitchondrial membranes, apparently in precisely linked assemblies that are repeated thousands of times within a single organelle. Some intermediate enzymes are dissolved in the mitochondrial matrix. In addition, many mitochondria contain the requisite enzyme systems for fatty acid oxidation, and they appear to participate in active and selective transport of water and certain solutes. Absorption and active extrusion of water occur, depending on the oxidation-reduction state of components of the respiratory chain on the mitochondrial membranes. The abundance of internal membranes may sometimes be correlated with the oxidative rate of the mitochondrion; in tissues such as mammalian liver where mitochondria serve many accessory functions, internal membrane area is small and matrix volume large. These facts give added interest to the association of mitochondria with the contractile vacuole (see below) in some protozoa, and suggest that *Pelomyxa*, with its occasional patterned mitochondrial structure, might be favorable material for further exploration.

Nearly every other cell component has been implicated as the source or site of development of mitochondria. In the protozoa these include small, undifferentiated, membrane-limited, precursor bodies arising *de novo* (*Paramecium*, Wohlfarth-Bottermann, 1958c) and nuclei (*Pelomyxa*, Brandt and Pappas, 1959; *Paramecium*, Ehret and Powers, 1955). In *Pelomyxa*, Brandt and Pappas, employing light microscopy, saw ordinary mitochondria adhering to the nuclear surface during and just after nuclear division. In electron micrographs the limiting membranes of the two organelles seemed to be continuous, and a filamentous material occurred in mitochondria that was similar to a substance layered under the nuclear membrane. An interesting sidelight on the question is provided by studies of the zooflagellates *Trypanosoma* (Steinert, 1960) and *Bodo* (Pitelka, 1961b). In the trypanosome, a mitochondrion is observed to be joined to the kinetoplast, an extranuclear, deoxyribonucleic acid (DNA)-containing body of unknown significance. In *Bodo* (Fig. 5, Pl. I), the cell's single, very long mitochondrion is continuous at both ends with the kinetoplast. In the absence of any developmental studies, it is not possible to conclude that mitochondria can originate at the nucleus, but temporary or permanent fusion, with opportunity for exchange of substances, between mitochondria and DNA-carrying bodies is conclusively demonstrated.

Chloroplasts

Chloroplasts, like mitochondria, provide the cell biologist with an organelle showing orderly and consistent membranous structure amenable to integrated morphological and biochemical analysis. If such studies involving electron microscopy at first seemed to lag because of difficulties encountered in the preservation of plant tissues, the weight and vigor of interest in photosynthesis nonetheless are pushing them to a high level of sophistication (see Calvin, 1959; Hodge, 1959; Wolken, 1959).

The pigmented phytoflagellates appear to occupy a gratifyingly intermediate position in a scale of increasing differentiation of chloroplast structure. In the photosynthetic bacteria, bacteriochlorophyll and carotenoids are localized in chromatophores that are minute (about 32 mμ in diameter) spheres with dense peripheries and low-density interiors (Bergeron, 1959). In the blue-green algae *Nostoc* (Hodge, 1959) and *Calothrix* (Ris and Singh, 1961), a membrane system ramifying through the cytoplasm is assumed to bear the enzymes that in eucells are localized in membranes of mitochondria and chloroplasts. In *Anabaena* and *Anacystis,* studied by the same authors, some of these membranes assume the form of wide, flat lamellae embedded in the general cytoplasm but often running in parallel below the cell membrane.

Primitive red algae (Brody and Vatter, 1959; Gibbs, 1960 and personal communication) have developed true chloroplasts, inasmuch as lamellar discs resembling those of the blue-greens now are segregated from the surrounding cytoplasm by two parallel limiting membranes. In all other algae (Gibbs, 1960) discs within the chloroplasts are very intimately associated in bands of two or more.

In various chrysomonad flagellates studied by Parke, Manton, and Clarke (1958, 1959), by Rouiller and Fauré-Fremiet (1958a), and by Manton and co-workers (see citations in Chapter 4) the lamellar bands run roughly parallel to the long axis of the chloroplast or to its curvature where it is cup-shaped. They converge at the two ends of the organelle, but some discs may extend less than its full length. The bands in these micrographs are rather sinuous and separated by spaces of varying width; to what extent this may be an artifact of fixation is uncertain.

Lamellae appear more regularly oriented in a dinoflagellate studied by Grell and Wohlfarth-Bottermann (1957) and in *Euglena,* examined by Wolken (1956, 1959, 1960), Gibbs (1960), and other workers (see citations in Chapter 4). Gibbs's detailed study shows occasionally as many as 12 discs in a band, but most commonly three. The bands are slightly sinuous on a small scale, but generally traverse the entire plastid in parallel arrays (Fig. 6, Pl. II), the width of the granular chloroplast matrix between bands being generally less than that of the bands. Where single discs end within the band or meet the limiting membrane of the plastid peripherally, a thickened rim is seen.

An interesting apparent variation of this lamellar pattern occurs in the chloroplast of the green algal flagellate *Chlamydomonas reinhardi* (Sager and Palade, 1957; Sager, 1959). Here the discs are imperfectly segregated into loose local stacks, which may be oriented at different angles within one plastid. Some discs may continue from one stack to another but many are restricted to a single stack. A similar arrangement of discontinuous bands, oriented circumferentially around a plastid that contains a large central or eccentric pyrenoid, is seen in three other green algae studied by Manton (1959a) and Manton and Parke (1960). This sort of arrangement suggests an approach to the much more rigidly ordered structure of the grana in chloroplasts of land plants. Lamellae in the latter may extend for some distance through the organelle, but local zones of compact piling, like stacks of coins, include some discs limited to one particular stack or granum, and account for most of the lamellar area at least in mature plastids.

In *Euglena* (Wolken, 1956), bleaching of chlorophyll to pheophytin is accompanied by breakdown of the lamellar structure. The organelles swell or collapse, depending on the treatment used to induce bleaching. Ghosts remain for some time, and upon renewed exposure to light under normal conditions, lamellar chloroplasts appear within a few hours. After prolonged periods during which *Euglena* survives saprophytically, chloroplasts are no longer recognizable in sectioned cells, and ultimately the cells are unable to regenerate chlorophyll in the light. DeDeken-Grenson (1960), however, working with centrifuged, dark-grown *Euglena* that were incapable of regenerating chlorophyll, found a particulate fraction containing yellow pigment. Sectioned pellets

PLATE II

FIG. 6. Section through chloroplasts of *Euglena gracilis,* showing bands of two to five closely appressed discs traversing plastids. Part of pyrenoid shown at bottom of left-hand chloroplast. At lower left, outside of plastid, is part of Golgi body; at right is edge of nucleus. × 30,000. From S. P. Gibbs, unpublished.

FIG. 7. Sectioned pyrenoid of *Chlamydomonas moewusii.* Dense pyrenoid matrix at center traversed by lamellae. Surrounding this is shell of low-density starch plates enclosed peripherally by chloroplast lamellae. × 15,500. From S. P. Gibbs, unpublished.

FIG. 8. Section of *Euglena gracilis* showing curved plate of dark granules constituting stigma and part of reservoir enclosing flagellum with paraflagellar body. × 33,000. From Gibbs, 1960.

FIG. 9. Longitudinal section through anterior end of *Ochromonas danica.* Two kinetosomes are sectioned (arrows), one of them continuous with its flagellum. Paraflagellar swelling extends toward stigma at upper end of chloroplast. This organism was fixed shortly after being returned to light following growth in dark; chloroplast is regenerating lamellae. × 41,000. From S. P. Gibbs, unpublished.

PLATE II

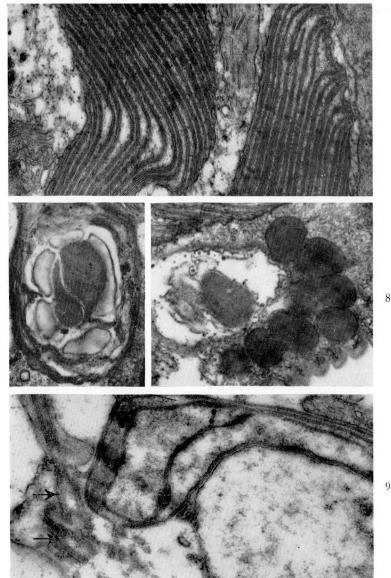

of this material contained loose, twisted, membranous bodies rather resembling mitochondria, which the author believed to be colorless plastid remnants permanently deprived of photosynthetic capacity. Similarly, Sager and Palade (1954) reported that a dark-grown yellow mutant of *Chlamydomonas,* which, unlike the usual form, cannot synthesize chlorophyll in the absence of light, contained reduced plastids with loose stigmas and no organized lamellae.

Thus, the integrity of lamellar structure appears to depend on chlorophyll synthesis. In higher plants, formation of chloroplasts from undifferentiated precursors occurs (von Wettstein, 1957), but algae appear to lack this capacity. Analysis of structure and pigment content by a variety of means has permitted Wolken (1956, 1959, 1960) to propose a molecular model for the *Euglena* chloroplast, with chlorophyll and carotenoid arranged in monolayers on the asymmetric lamellar membrane.

Gibbs's (1960) analysis of *Euglena* clarifies the structure of the pyrenoid, a specialized region of the chloroplast in many algae that is believed to function in the elaboration or concentration of reserve products of photosynthesis. Such products usually are found as grains or shells around the pyrenoid. The pyrenoid area of the *Euglena* plastid has a denser matrix than the rest of the chloroplast although continuous with the latter (Fig. 6, Pl. II). It evidently is traversed by all adjacent lamellar bands, but each band is reduced to two (rarely one or three) discs as it enters the pyrenoid. Within the pyrenoid, the spacing between the narrow bands is more constant than elsewhere. Pyrenoids in the chryso-monads and phytomonads (Fig. 7, Pl. II) similarly consist of discrete regions of denser matrix typically traversed by only a few, rather widely separated, bands that may consist of only a single disc, or by more or less contorted tubules which, however, are continuous with adjacent lamellae.

An eyespot or stigma occurs in many pigmented phytoflagellates as a small reddish body functioning in photoreception. The pigment is a carotenoid identified in some species as astaxanthin, a β–carotene derivative. In *Chromulina psammobia* (Rouiller and Fauré-Fremiet, 1958a) and *Chlamydomonas reinhardi* (Sager and Palade, 1957) it occupies a differentiated region of the chloroplast,

c

while in *Euglena gracilis* (Wolken, 1956, 1959, 1960; Gibbs, 1960) it is a separate body (Fig. 8, Pl. II). In all of these instances, it consists of a number of dense granules, apparently membrane-limited, arranged in one to three flat or curved plates. The granules are up to 400 mμ in diameter and show in some aspects a rather regular hexagonal packing.

The stigma is considered in most organisms to be a primary light receptor, but in *Euglena* it has been thought to function as a light-absorbing shield that permits directed phototropic response by intermittently shading the true photoreceptor, a swelling on the adjacent flagellum (Fig. 8, Pl. II). This swelling or para-flagellar body consists of a dense, homogeneous mass bordered by a light space and surrounded by the flagellar membrane (Roth, 1958a; de Haller, 1959, 1960; Gibbs, 1960). It is not known to contain any photosensitive pigment; hence, as Gibbs points out, its function as a direct light receptor is extremely dubious.

Whether or not the paraflagellar body is directly sensitive to light, it is highly significant that a stigma-flagellum association has been detected in other flagellated algal cells by recent electron-microscope studies. Thus in the spermatozoid of *Fucus*, Manton and Clarke (1956) found that the recurrent flagellum is swollen and adherent to the cell surface over the length of the peripherally located eyespot, becoming free and cylindrical thereafter. In *Ochromonas danica* (Fig. 9, Pl. II) a paraflagellar swelling similar to that of *Euglena* is seen adjacent to the stigma (Gibbs, unpublished). The most intriguing example yet is that of *Chromulina psammobia,* previously supposed to be uniflagellate, where Rouiller and Fauré-Fremiet (1958a) found a short, swollen, but structurally impeccable, flagellum entirely enclosed within a cylindrical invagination of the cell surface and occupying the groove formed by the curved plate of the stigma. Here is a flagellum, clearly not active in locomotion, whose only obvious association is with a light-sensitive organelle. Many examples of cilium-associated sensory structures in metazoa come to mind, notably the retinal receptor cells of vertebrates. In these, the inflated outer segment of the rod or cone is packed with pigment-bearing lamellae and is connected to the inner segment, and hence to the sensory nerve fiber, only by a short, slender, ciliary stalk (p.45).

Contractile Vacuoles

Contractile vacuoles are by no means peculiar to the protozoa, but they are so typical of familiar fresh-water species that we tend to think of them as a protozoan property, if not monopoly. They are absent in many marine and symbiotic forms, and this fact early gave rise to the logical assumption that they are instruments of osmoregulation. This is supported to a large extent by experimental data (Grassé, 1952; Kitching, 1956a; Fauré-Fremiet and Rouiller, 1959). The contractile vacuole appears and grows by the confluence of smaller droplets. Periodically its contents are discharged through the cell surface. The amount of fluid that it pumps out of the cell under normal conditions may within a period of ten minutes equal the total water content of the cell (Fauré-Fremiet and Rouiller, 1959). In most species tested, the rate of cyclic contraction and/or the maximum size of the vacuole before discharge vary inversely, within limits, with the concentration of solutes in the external medium; whether this is a direct effect on the vacuole and its filling mechanism or an indirect one modifying permeability of the cell membrane is not known.

In addition to an osmoregulatory function, a role in the secretion of metabolic wastes has been suggested, but pertinent experimental data are lacking. Whatever its ancillary functions, the sheer volume of water (derived from ingestion of water with food and from diffusion through the cell surface) transported through the cell by this means necessitates some mechanism for the selective retention or resorption of protoplasmic solutes. The consensus among recent students of the contractile vacuole is that an active phase segregation or secretion most probably occurs in the zone where the first contributory vesicles appear. Analogy with the nephridial organs of metazoa is clearly suggested. Perhaps the analogy should not be carried too far, as the water-conserving function of renal organs in terrestrial or marine animals is not known to have any parallel in the activities of the contractile vacuole; however, the activity of such vacuoles in certain internal symbiotic protozoa needs to be studied.

Because of its accessibility, the protozoan contractile vacuole offers an elegant experimental material, still largely unexploited, to investigators concerned with intracellular water transport mechanisms, and a considerable amount of information already available on its fine structure provides an added attraction.

In amebae, contractile vacuoles move about through the protoplasm and presumably have no morphologically determined discharge pore. Light microscopy indicates a rather thick vacuolar membrane and surrounding this a layer of gelated protoplasm and a cloud of so-called beta-granules. In the electron microscope the vacuole in *Amoeba proteus, Pelomyxa carolinensis,* and *Hartmannella rhysodes* (Greider, Kostir, and Frajola, 1958; Pappas, 1959; Mercer, 1959) is limited by a typical unit membrane. The protoplasm surrounding it is filled with tubules and vesicles, 20 to 200 mμ in diameter. This vesicular zone varies from 0·5 to 2 μ in thickness, probably depending on the stage in the vacuolar cycle. Beyond it is a halo of mitochondria (the beta-granules of the light microscopist), irregularly crowded while the growing vacuole is small but becoming aligned in a compact ring as the vacuole enlarges. Rather frequently, individual vesicles appear to open into the vacuole, and Mercer and Pappas agree in their interpretation of this picture as one suggesting segregation of fluid into membrane-bounded vesicles and tubules, and the emptying of this fluid into the main vacuole by coalescence. The mitochondria presumably provide energy for the segregation, and may in addition be involved in the water transport itself. Vacuoles at the moment of systole in amebae have not been shown.

A canalicular-vesicular zone surrounding the vacuole has been found in several ciliates, including the suctorian *Tokophrya* (Rudzinska, 1958), the peritrichs *Campanella* and *Ophrydium* (Fauré-Fremiet and Rouiller, 1959), and several astomes (de Puytorac, 1960, 1961a). In *Tokophrya* the vesicles do not appear to be particularly numerous; in the peritrichs and the astomes, on the other hand, the thick cortical zone is permeated by distinctly tubular elements that branch and anastomose and occasionally dilate to form larger vesicles. In all cases, occasional images suggest the opening of tubules or vesicles into the main vacuole. Around the periphery of the spongy zone, membranes of the tubules occasionally appear to be continuous with membranes of the endoplasmic reticulum. Mitochondria are variably abundant in the surrounding cytoplasm.

Relatively long discharge canals lead from the vacuole to the exterior in the two peritrichs; these have thick (65 mμ), apparently homogeneous walls and appear during diastole to be closed at both proximal and distal ends by membranes. In *Tokophrya* (Fig. 10, Pl. III), a permanently open external channel leads to the body surface from a small papilla projecting into the contractile vacuole; in the papilla the lumen of the canal narrows to a very fine tubule during diastole but is widely expanded during systole. Fine fibrils, about 18 mμ in diameter, radiate from the channel to the adjacent vacuole membrane and might, if they are contractile, be responsible for changes in the diameter of the excurrent tubule.

In the astome *Metaradiophrya gigas* the discharge canal consists of a distal invagination of the cell surface, with annular fibers in one side of its wall and a proximal cone-shaped part ending in a papilla in the main vacuole wall. This proximal end is closed off by a septum from the vacuole cavity. Radially arranged fibrils that look tubular in section pass from the canal wall to the vacuole's cortex (Fig. 11, Pl. III). Some other astomes have a contractile vacuole apparatus that is a permanent long tube. This undergoes rhythmic waves of contraction, and discharges by a series of well-marked pores. The latter have longitudinal fibers in their walls and, again, radial fibers leading from their proximal parts toward the main vacuole wall.

Contractile vacuoles of an unusual sort have been observed by Noirot-Timothée (1960) in several entodiniomorph ciliates. These intestinal symbiotes of ungulates have vacuoles that pulsate at rather infrequent intervals, with a prolonged resting stage interrupting diastole. Light microscopy shows a basophilic and osmiophilic cortex about the vacuoles. In electron micrographs this zone during the resting stage appears as a sparse halo of tiny, clear, spherical vesicles. In some micrographs, scraps of ergastoplasm are present in the adjacent cytoplasm.

The most complex contractile vacuole apparatus yet examined is that of *Paramecium*. In most species of this genus the contractile vacuole is a reservoir fed by pulsating radial canals. Schneider (1960a, 1960b) in a thorough electron-microscope study found in *P. aurelia* and *P. caudatum* the apparatus diagrammed in Text-fig. 2. A network of fine canaliculi, 15–20 mμ in diameter, forms a cylindrical field around each radial canal (Fig. 12, Pl. III). At the periphery of the field these nephridial tubules appear in places to

PLATE III

FIG. 10. Section through contractile vacuole canal of *Tokophrya infusionum*. Outer, permanently open, part of canal is obliquely cut above; below, contracted tubule leading through papilla from main vacuole is cut tangentially. Note fibrils radiating from canal to vacuole wall. × 50,000. From Rudzinska, 1958.

FIG. 11. Oblique section through part of contractile vacuole canal (cavity at upper right) of *Metaradiophrya gigas*. Cylindrical fibrils are regularly arranged along one side of canal wall; some fibrils radiate to membrane of contractile vacuole, at left. × 47,000. From de Puytorac, 1961a.

FIG. 12. Oblique section through radial canal of *Paramecium caudatum*. Canal is collapsed in systole, as in upper circle in Text-fig. 2. Nephridial tubules form sponge around canal; clusters of tubular elements and canaliculi of endoplasmic reticulum visible in surrounding cytoplasm. × 34,000. From Schneider, 1960b.

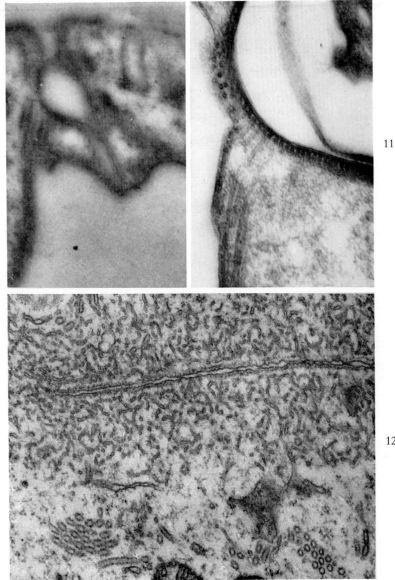

Plate III

10

11

12

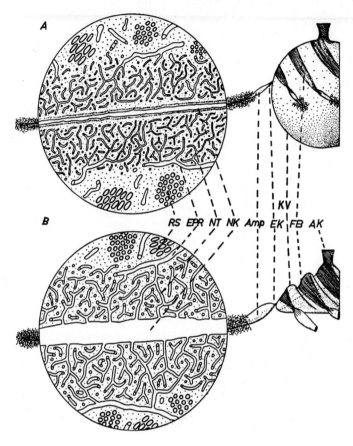

TEXT-FIGURE 2. Schematic drawings of the contractile vacuole apparatus in *Paramecium* showing main vacuole to right and one radial canal, with enlarged circular inset, to left. Upper drawing shows radial canal in systole, main vacuole in diastole; lower drawing shows radial canal in diastole and main vacuole in systole. RS, clusters of membranous tubules; EPR, endoplasmic reticulum, showing continuity of this system with NT, nephridial tubules forming a sponge around NK, nephridial canal; Amp, ampulla of radial canal; EK, injector canal; KV, main contractile vacuole; FB, bundles of fibrils in vacuole wall; AK, discharge canal. From Schneider, 1960b.

be continuous with endoplasmic reticulum. The radial canal has a round cross-section in diastole and apparently collapses like a balloon at systole. At this time, a narrow layer of homogeneous substance is present between the walls of the deflated radial canal and the adjacent tubules, but during diastole this layer disappears and some of the tubules appear to open into the canal. Around the swollen medial ends, or ampullae, of the radial canals, the surrounding sponge of nephridial tubules is less dense. A narrow injector canal, around $0.5\ \mu$ long, leads from each ampulla obliquely into the main contractile vacuole; nephridial tubules are lacking around the injector canal and around the vacuole. Fine fibrils, about 20 mμ in diameter and appearing tubular in cross-section, are seen singly on the outer surface of the membrane of the ampullae. They pass along the walls of the injector canals, joining to form increasingly wider ribbons, and continue, in tracts of ten to 40 fibrils each, along the outer, or pellicle, side of the contractile vacuole. The vacuole itself has a membrane indistinguishable from that of the radial and injector canals. In diastole, the vacuole is round; in systole the inner wall of the vacuole is flattened but smooth while the outer wall is deeply folded in regular waves, with the bands of fibrils occurring on one slope of each wave. The fibril bands continue and describe a spiral about the discharge canal, which is formed of invaginated cell membrane. It is open externally and, during diastole of the vacuole, closed at its inner end by a double septum consisting of cell membrane on the outside and vacuole membrane on the inside.

In the cytoplasm surrounding the zone of nephridial tubules are mitochondria in moderate numbers and clusters of tube-like elements (Fig. 12, Pl. III) of unknown significance. Individual tubes are about 50 mμ in diameter and are usually packed in orderly rows within the cluster.

This picture suggests several correlations with the observable events of the contractile vacuole cycle. The cycle involves first the filling of the radial canals while the main vacuole remains collapsed; this would require an effective closure of the injector canals. The ampullae expand considerably more than the elongate radial canals; conceivably the spongy tubular zone surrounding the latter may resist stretching. After maximal filling, the radial canals and ampullae all empty rather rapidly through the reopened injector canals as the main vacuole expands. Schneider suggests that systole of the main vacuole is initiated by contraction of the fibers seen in its walls. These would first close the injector canals and then exert pressure on the vacuole contents, forcing rupture of the septum covering the exit pore. Upon release of its contents, the medial wall of the vacuole is pressed toward the lumen, possibly by elastic reaction of the surrounding cytoplasm to compression suffered during diastole. The septum over the exit canal is repaired during the resting phase following systole, while the radial canals are filling and the main vacuole remains collapsed.

This hypothesis would appear to be mechanically plausible. The arrangement of the 20-mμ fibrils along the walls of the system is consistent with the assumption that they are contractile, and similar fibrils have been observed in other contractile organelles among the protozoa (see descriptions of *Pyrsonympha,* Chapter 5, and *Stentor,* Chapter 6), as well as in association with contractile vacuole pores of other ciliates described above.

If, as several studies indicate, a membranous septum closes the exit canal and must undergo rupture and repair with each pulsating cycle (that is, every few seconds), this constitutes a rather remarkable example of membrane activity.

In all of these studies, the cytoplasmic zone in which segregation of water and resorption of solutes almost certainly takes place is seen to be occupied by tubules or vesicles providing a large, and in the more extreme cases enormous, membrane surface area. Linkage of these tubules with endoplasmic reticulum may be significant if the latter system can be demonstrated to function in water transport elsewhere in the cell; the contractile vacuole apparatus occupies, after all, only a relatively small part of the total cell volume. In fact, this picture, as Schneider points out, constitutes one of the best parcels of morphological evidence that such may be the case. Needless to say, proof of any of these assumptions will require more than morphological evidence alone. It would be interesting to learn whether contractile vacuoles with a conspicuous cortex of packed tubules are more active in their output than those with relatively fewer surrounding vesicles.

One question that is particularly puzzling is how the vacuoles or radial canals that receive water from the surrounding sponge swell as much as they do. The influx of fluid during diastole must create enough hydrostatic pressure to displace the surrounding cytoplasmic mass, yet fluid continues to enter.

Contractile vacuoles in flagellates have been seen occasionally in electron micrographs; in several instances, simple membranes with no apparent cortical differentiations have been described. This is true of several phytoflagellates where contractile vacuoles are conspicuous in life, and a highly differentiated cortex, if present, would surely have been observed by the investigators. Gatenby, Dalton, and Felix (1955) have pictured a lamellar body resembling the conventional Golgi apparatus bordering the contractile vacuole of some flagellates, and they cite this in support of Nassanov's hypothesis (disputed by Grassé and Hollande, 1941, among others) of the homology of protozoan contractile vacuoles with the Golgi complex.

The clearest micrographs of flagellate contractile vacuoles are those of *Paraphysomonas vestita* by Manton and Leedale (1961c). The bounding membrane is somewhat convoluted in the pictures shown, and around it are many smaller vacuoles, but no layer of packed convoluted tubules. The Golgi body lies nearby, but its closest edge appears to be nearly a micron away from the vacuole. Golgi microvesicles are similar to the small vesicles surrounding the contractile vacuole, so a migration from the Golgi zone is not inconceivable, but neither is it strongly supported by the available evidence. In many flagellates Golgi and contractile vacuole both occupy fixed positions near the kinetosomes, but evidence to suggest a direct functional relationship is lacking.

Pinocytosis and Phagocytosis

The phenomena of phagocytosis and pinocytosis are treated together because current evidence suggests that similar processes are involved, even in species that have permanent mouth openings.

Pinocytosis or " cell drinking " was first described (for historical citations and review, see Holter, 1959) by Lewis in 1931 in cells in tissue culture. Mast and Doyle in 1934 identified a similar process in amebae, and in recent years the phenomenon has attracted wide interest. At the light-microscopic level, pinocytosis in amebae typically involves the formation of many slender channels from the cell surface deep into the cytoplasm. The channels become constricted as strings of vesicles which may then decrease in size. Instead of narrow channels, initial pinocytotic invaginations may be wider, even as wide as food cups, so that the only distinction between the two processes appears to be the indeterminate one of food size — solutions or macromolecular suspensions in the case of pinocytosis versus microscopically visible particulate ingesta in phagocytosis. Intense pinocytotic activity is stimulated in amebae by the presence of certain substances in the medium, notably proteins. The ultimate fate of pinocytosis vesicles is unknown, and this is currently an important mystery.

Electron microscopy discloses apparently similar processes on a more minute scale in many vertebrate cells as well as in protozoa. Thus the engulfment of a bit of cell membrane surrounding a droplet of ambient fluid occurs in a wide range of cells, and the vesicles so formed vary widely in size.

All of this has assumed particular interest in view of the known absorption by living cells of molecules presumably too large to diffuse across the cell membrane. Pinocytosis is an attractive possibility as a partial explanation of this phenomenon, but the fact remains that the contents of the pinocytosis vesicle are still separated by a membrane from the cytoplasm. Does the membrane rupture or decompose in the cytoplasm to release its contents? Is the membrane permeability altered after engulfment? Does digestion occur within pinocytosis vesicles, permitting subsequent diffusion of the products? Or does the increased membrane area resulting from intense pinocytosis merely augment the total quantity passed of a slowly diffusing substance? In the case of phagocytosis, digestion within a food vacuole has long been believed to occur; morphologic changes in the ingested food can be observed microscopically, and changes in the pH of the vacuole contents are clearly demonstrable. If similar processes take place in pinocytosis vesicles, then it would seem necessary to conclude that any bit of cell membrane is a potential digestive surface when surrounded by any part of the cytoplasm. However, it may be significant that the hairy plasmalemma coat of *Amoeba* seems to persist for some time on pinocytosis vesicles or deep channels, whereas it disappears very quickly (see below) from food vacuole membranes. Opinions on these matters vary so much that at present writing the questions cannot be answered, but vigorous studies of pinocytosis in amebae are under way in many laboratories (*e.g.*, Holter, 1959; Brandt, 1958; Schumaker, 1958).

Using fluorescent or radioactive-labeled proteins, investigators have found that amebae are capable of taking up impressively large quantities of protein from the medium, far more than could be accounted for by simple ingestion of fluid. Apparently the protein is adsorbed to the plasmalemma, and Brandt (1958) suggests that surface binding weakens the membrane tension at points which are consequently drawn into the cell by the contractile plasmagel system. Fluorescent proteins are visible some days after uptake as multitudinous small fluorescing granules whose changing densities in cells stratified by centrifugation suggest that digestion is taking place. Radioactive glucose, administered with unlabeled protein as a pinocytosis inducer, rapidly results in general cytoplasmic labeling, showing absorption of glucose or its breakdown products, although the plasmalemma normally is nearly impermeable to glucose itself. The label also was recovered from respiratory CO_2 and excretion products.

While the above work is based on light-microscope studies, several recent reports by electron microscopists illustrate the morphology of some steps in the process of pinocytosis in amebae. Brandt and Pappas (1960) used colloidal suspensions of thorium dioxide particles and solutions of ferritin as both pinocytosis inducers and electron-microscopically visible labels. The dense particles of the label in both instances became concentrated on the surfaces of fine hairs that typically coat the plasmalemma of amebae (see Chapter 3). Particularly with thorium dioxide, the hairy fringe became thicker and was densely impregnated with particles; ferritin tended to concentrate in a layer at the tips of the hairs. Convoluted tunnels extending deep into the cytoplasm were lined by the impregnated fringe. Subsequent phases in the history of the tunnels were not described in this first publication, except for the notation that tiny vesicles containing densely concentrated particles were observed some hours after exposure to thorium doxide. In any event, morphological evidence now clearly confirms the hypothesis that quantitatively significant adsorption of dissolved or suspended substances occurs at the plasmalemma surface.

Roth (1960a), studying ultrastructure of normal, untreated amebae, observed vacuoles well within the cytoplasm with membranes bearing the fine plasmalemma fringe. Other vesicles lacked the fringe; some had dense granules on their membranes; some vesicle membranes were darkened by staining with phosphotungstic acid, while others were not. Roth tentatively suggested a sequence of changes in membrane properties, including permeability. Demonstration of such a sequence will require the use of labels that are capable of entering the cytoplasm of the ameba.

Chapman-Andreson and Nilsson (1960) were unable to see any difference in the appearance of pinocytotic and other intracellular membranes and the plasmalemma, but their micrographs were not sufficiently detailed to reveal membrane structure. They noted the presence of small vesicles in the region of pinocytosis channels, similar to those surrounding the contractile vacuole, and speculated that these might be associated with water removal from the channels.

Wohlfarth-Bottermann (1960b), in an electron-microscope study of the small free-living ameba, *Hyalodiscus,* observed abundant vesicles, especially at the posterior end of the cell, that still bore the hairy plasmalemma fringe and thus could be identified as pinocytosis channels or vacuoles. Other vesicles within the endoplasm had no fringe but a distinct three-ply structure (Fig. 13, Pl. IV). Still others, apparently sectioned at similar angles and often continuous with three-ply membranes, appeared to be limited by simple single-ply membranes. Furthermore, some of the three-ply membranes were fragmented, suggesting release of vesicle contents into the cytoplasm. Usually, fragmented membranes are attributable to polymerization damage in the plastic blocks, but Wohlfarth-Bottermann's specimens were embedded in Vestopal, which presumably reduces polymerization damage to a minimum, and the general appearance of his sections indicated superior fixation for ameba protoplasm, as well as high-resolution micrography. Taken at face value, his results not only suggest a sort of membrane fragmentation not certainly seen elsewhere, but also cast some doubt on the unit-membrane concept. The use of labeled proteins as pinocytosis inducers with *Hyalodiscus* might yield highly significant information.

Amebae again provide the bulk of current information on the

PLATE IV

FIG. 13.   Section of *Hyalodiscus simplex,* showing cytoplasmic structure and inclusions. Note variations in density of cytoplasmic matrix in inclusion-free ectoplasm. Vesicles limited by unit membranes visible in cytoplasm; one marked by arrow has fragmented membrane. Numerous, opaque particles of various sizes are characteristic of most amebae; their significance is unknown. × 46,600. From Wohlfarth-Bottermann, 1960b.

FIG. 14.   Tangential section through cortex of minute vesicles surrounding digestive vacuole of *Pelomyxa carolinensis.* × 7,500. From Roth, 1960a.

FIG. 15.   Cross-sections through several kinetosomes of the complex zooflagellate, *Barbulanympha* sp., showing skewed triplet fibrils, cartwheel arrangement of filaments in lumina, and pattern of filaments around and between kinetosomes. × 60,400. From J. E. Cook, unpublished.

FIG. 16.   Section of nucleus of *Bodo saltans,* showing double envelope with pore at left center; cytoplasmic extensions of outer envelope layer at arrows. Central, coarsely granular nucleolus; peripheral, finely granular chromatin. × 56,000.

PLATE IV

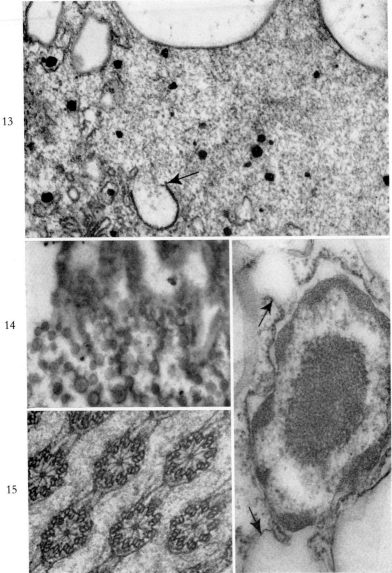

morphology of phagocytosis. Roth's (1960a) study confirms and considerably extends the observations of several previous electron microscopists (see section on amebae in Chapter 3). Food vacuoles in protozoa soon after formation typically shrink by elimination of ingested water, simultaneously becoming acid as the prey is killed (Kitching, 1956b). Subsequent alkalinization occurs, with some increase in volume, and ultimately the food is reduced to an undigested residue surrounded by fluid; the vacuole contents then are expelled through the cell surface. Correlating his electron-microscope studies on *Pelomyxa* with light-microscope observations by Mast, Roth identifies the following: (1) Recently formed vacuoles containing unchanged food organisms and limited water in a smooth, thin membrane from which the plasmalemma fringe already has disappeared. (2) Early stages of loss of structure of the food organism, when the vacuole membrane shows many long villous protrusions and foldings toward the surface of the prey. Roth believes that this appearance coincides with the addition of alkaline substances to the vacuole from the ameba's protoplasm. (3) Vacuoles found abundantly during the longer phases of digestion and absorption of prey. A homogeneous substance is layered against the vacuole side of the membrane, and the latter, irregular in contour, is surrounded by a dense cloud of tiny vesicles containing similar material (Fig. 14, Pl. IV). Budding of these vesicles from the membrane is suggested in several micrographs, and Roth concludes that transfer of vacuole contents into the cytoplasm by micropinocytosis is occurring. No means of tracing the pinocytosis vesicles after their formation was available. Roth emphasizes the enormous increase in membrane area provided by such a process. If most of the contents of a food vacuole are thus parceled out during a period of 12 hours, it would be necessary that each μ^2 of original vacuole surface be duplicated every minute. Interestingly, no special cytoplasmic organelles such as mitochondria or Golgi membranes appear regularly in the vicinity of the vacuole at any stage.

A similar process of pinocytosis around phagocytosis vacuoles is indicated in electron micrographs of the heliozoan, *Actinosphaerium,* studied by Anderson and Beams (1960). Jurand (1961), in a study of food vacuoles in *Paramecium,* noted similar but less abundant small vesicles surrounding food vacuoles presumably in a digestive phase; again, his micrographs suggested that the vesicles were forming by a budding process from the vacuole membrane. In addition, Jurand observed minute invaginations of the plasma membrane at the base of the oral cavity, near the site of food vacuole formation. Möller and Röhlich (1961) examined food vacuoles in *Tetrahymena corlissi* and again found evidence of pinocytosis from the main vacuole during digestive phases. Ingested cells were unrecognizable from the first, suggesting some extracellular predigestion in the buccal cavity.

Flagellar Apparatus

Cilia and flagella were among the first objects of electron-microscope study by biologists, for the obvious reason that, until thin-sectioning techniques were developed, they were almost the only protoplasmic structures in organisms above the bacteria that could be examined without resorting to cell fragmentation. A number of early investigators turned their attention to animal sperm tails (see reviews by Fawcett, 1958, 1961) and to the flagella of protozoa and metazoa (Schmitt, Hall, and Jakus, 1943; Brown, 1945; Jakus and Hall, 1946; Dinichert, Guyénot, and Zalokar, 1947; Foster, Baylor, Meinkoth, and Clark, 1947; Pitelka, 1949). The consensus of all these studies, now mainly of historical interest, was that the axis of these vibratile organelles consisted of a number, usually described as nine to 12, of continuous longitudinal fibers, and that these were bound together by a sheath of variable composition and dimensions. In addition, Brown (1945) and Pitelka (1949) demonstrated the presence in several flagellate species of orderly arrays of lateral filaments. None of these observations was totally new, since light micro-scopists (see papers by Brown and Pitelka for references) had detected fibrous axial and filamentous lateral structures.

In the early 1950's the English botanist Manton and her colleagues, using an ingenious combination of techniques for the electron-microscope study of whole and fragmented small plant flagellates, came to the startling conclusion that the axis in all cases contained precisely 11 longitudinal fibers, and that these were always arranged as a cylinder of 9 fibers about a central pair. The latter, finer than the others and more likely to disintegrate during preparation, presumably had a somewhat different composition. Manton predicted that the same arrangement would be shown to obtain in other flagella. In 1953, Fawcett and Porter published the first high-quality electron micrographs of thin-sectioned cilia, from a variety of metazoan ciliated epithelia, and confirmed Manton's prediction for this material. Subsequent studies by Manton and by a multitude of other workers enable us to state as a general truth that all motile cilia and flagella (above the bacteria)* are organized on the 9+2 pattern.

A number of studies of frayed flagella (Manton and Clarke, 1952; Pitelka and Schooley, 1955) seemed to show that the 11 fibers of the axis were themselves multiple structures, being composed of up to five or even more subfilaments. Examination of sectioned flagella consistently belies this appearance. The nine peripheral fibers are in all cases double, while the central ones are single. Recent high-resolution micrography has revealed a considerably greater structural complexity than the earlier images showed. The most detailed information for protozoa comes from the superb high-resolution study by Gibbons and Grimstone (1960) on several species of hypermastigote flagellates from the termite gut.

In *Trichonympha* (and the other organisms studied by Gibbons and Grimstone) the central fibrils of the flagellum are about 24 mμ in diameter and about 30 mμ apart, center to center (Text-fig. 3). They may be enveloped in a common sheath. Like the peripheral fibrils, they appear tubular in section. This does not necessarily mean that they are hollow, but only that their margins are of higher density than their cores. Some pictures suggest a fine periodic banding in the central fibrils, perhaps indicating that each is composed of a two- or three-strand helix. The 8-shaped

---

* Unfortunately, there is no term in common English usage that includes cilia and flagella and excludes the much finer, filamentous appendages of bacteria, which certainly are not homologues of the 9+2 flagella. Dougherty (1957b) has proposed the neologisms, *proterokont* for the bacterial appendage and *pecilokont* for the flagella and cilia of all higher organisms. These words are eminently useful when both groups of organisms are under discussion. Through most of the present review, proterokonts will be excluded from consideration, and *flagellum* will be used as an inclusive term; cilia are merely one of several varieties of flagella.

D

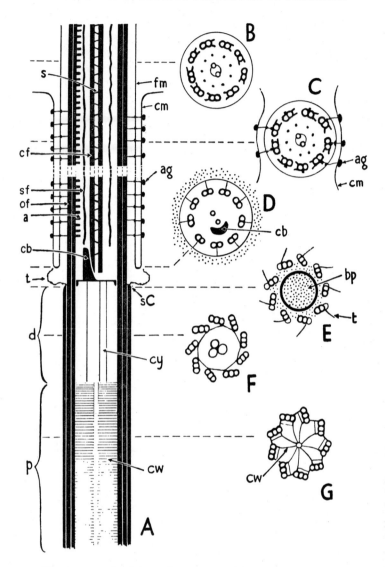

TEXT-FIGURE 3. Schematic representations of a flagellum and kinetosome in the anterior body region of *Pseudotrichonympha*. A shows longitudinal section, B–G are cross-sections at the levels indicated. a, lateral arms on the peripheral fibrils of the flagella; ag, granules beneath the cell membrane in the wall of the groove double peripheral fibrils are about 25 by 37 mμ in cross-section and average about 55 mμ apart, center to center. Each one is so placed that one end of the figure 8 (one of the subfibrils composing it) is very slightly closer to the center of the flagellum than the other. From the closer subfibril, a pair of short arms, about 5 by 14 mμ in cross-section and less dense than the outline of the fibril itself, extends toward the next adjacent fibril. These always point in the same direction on all fibrils in a flagellum, and in three zooflagellates, a ciliate, and several metazoan epithelia (Gibbons, 1961) this appears to be clockwise when viewed from the basal end of the flagellum. In longitudinal sections the arms are indistinct, but apparently they are discontinuous projections. Both central and peripheral fibrils are straight; they do not spiral or twist except as the whole flagellum does, and they maintain a strikingly constant spatial relationship within the bundle.

Nine very slender secondary filaments occur midway between the outer fibrils and the central pair. These are about 5 mμ in diameter and rather sinuous on a small scale; they usually are inconspicuous and rarely are detectable in longitudinal sections. The whole complex of fibrils forms a cylinder somewhat under 0·2 μ in diameter; this measurement may vary in different species. The fibrils are embedded in a matrix of low density and surrounded by a unit membrane that is continuous with the plasma membrane of the cell. Often this bears irregular, minute, villous projections.

At their tips, most flagella taper more or less abruptly. In *Trichonympha* the diameter of the fiber bundle decreases first while still retaining the 9+2 pattern (see also Roth, 1956); the lateral arms on the peripheral fibrils and the nine secondary filaments disappear at this level, which suggests that their positions are in

---

in which flagellum lies; these are connected by minute bridges to smaller particles under the membrane of the flagellum; bp, plate or septum across top of kinetosome; cb, crescent-shaped, asymmetric axial granule; cf, central fibril of flagellum; cm, cell membrane; cw, cartwheel structure occupying basal part of kinetosome; cy, cylinders in lumen of distal part of kinetosome; d, distal part of kinetosome; fm, flagellar membrane; of, peripheral fibril of flagellum; p, proximal region of kinetosome; s, sheath surrounding central fibrils; sC, distal end of third subfibril of kinetosomal fibril; sf, secondary fibril; t, fine filament extending laterally from apex of kinetosome. From Gibbons and Grimstone, 1960.

some manner related to the maintenance of fixed radial and circumferential distances in the complete bundle. Farther distally the axial fibrils lose their regular spacing, subfibrils of the peripheral nine end at random levels so that the double profiles give way to single ones, and the total number visible in cross-section diminishes to zero. The plasma membrane covers the tip of the flagellum.

A quite different kind of flagellar tip is present in the very tiny green flagellate *Micromonas pusilla* studied by Manton (1959a). Here the flagellum proper is reduced to a stub no more than 1 μ in length, while a much longer distal thread consists of a prolongation of the two central fibrils only, still surrounded by the flagellar membrane.

It seems probable that Gibbons and Grimstone's description will prove to be applicable to all flagella and cilia. Similar components are evident in micrographs of the sperm tails of an insect by André (1961), of sea urchin sperm tails and ctenophore swimming-plate cilia by Afzelius (1959, 1961), and of rotifer coronal cilia by Lansing and Lamy (1961). Published micrographs of other species and unpublished micrographs from the author's laboratory indicate that similar structures occur widely, but few other materials are so favorable for detailed study as the hypermastigote flagellates with their neatly aligned, close-packed flagella, and few other authors have achieved such elegant preservation and micrography as Gibbons and Grimstone.

Cilia, whether of ciliated protozoa or of multicellular organisms, tend to show minimum elaboration of the basic design just described (although some unusual structures are present in long, close-packed cilia of the swimming plates of ctenophores, as shown by Afzelius, 1961; these probably are peculiar to these extraordinary organs, since other cilia of the same organism are more conventional). Flagella of flagellates frequently, but by no means always, have expanded sheaths, often enclosing rods or ribbons of peculiar structure, surrounding the conventional fibrous axis. In addition, fine filamentous lateral appendages or mastigonemes occur in characteristic patterns on the flagella of many phytoflagellates. Since these features vary widely and tend to be consistent within particular taxonomic groups, they cannot be requisite to flagellar motility, and they will be discussed in connection with the species concerned. Interesting peripheral structures, often paralleling the nine outer fibers of the axis, are found in many animal spermatozoa (see Fawcett, 1958, 1961).

The basal regions of flagella and their intracellular parts, the basal bodies or kinetosomes, differ somewhat in the various groups of organisms studied. Additional fibrous structures of diverse pattern and construction are attached to kinetosomes in different species, and will be discussed later. But a definite and highly significant pattern is common to the morphology of kinetosomes in general.

The kinetosome consists of a cylinder of nine fibrils that continue distally as the nine peripheral fibrils of the flagellum (Text-fig. 3). Evidence from some pioneer papers (e.g., Fawcett and Porter, 1954; Bradfield, 1955) indicates that a short interruption may occur in these nine fibrils where they emerge from the cell surface in certain metazoan ciliated epithelia, but this work needs to be repeated using more recent techniques. For the most part, the continuity is unquestionable. The kinetosome almost always is short, about 0·5 μ, but in hypermastigote flagellates such as *Trichonympha* it may extend up to 4 μ in length. At least in the proximal part of the kinetosome the nine fibrils typically are not double, but triple, and the row of three is skewed from the circumferential direction somewhat more than is the doublet in the flagellum (Fig. 15, Pl. IV). This probably accounts for an apparent thickening of the proximal parts of the fibrils as seen in longitudinal section. These features have been noted several times (e.g., Rhodin and Dalhamn, 1956; Noirot-Timothée, 1959; Afzelius, 1959; Gibbons and Grimstone, 1960) and are visible by hindsight in many other published micrographs.

In metazoan ciliated epithelia, kinetosomal fibrils generally converge basally and mingle in a dense matrix material; the kinetosome thus is closed off at its tapering inner end. Protozoan kinetosomes with few exceptions are open proximally and maintain a nearly constant diameter throughout. Within the low-density core of the kinetosome, rods, granules, radial fibrils, or tubular structures may appear, or the entire interior may look empty. A cartwheel-like configuration in the center of cross-sectioned kinetosomes is not uncommon. Regularly arranged fine filaments connecting specific subfibrils of the nine fibrils with each other or with the center of the kinetosome, or interconnecting adjacent kinetosomes, are described in *Trichonympha* by Gibbons and Grimstone.

The distal end of the kinetosome is a zone of particular interest. Here the two central fibrils of the flagellum originate, most frequently at a small dense body that may be spheroidal, flat, or markedly asymmetric. Below the granule, or at the same level if no granule is present, one or more thin septa may stretch partly or completely across the core. The distal terminus of the kinetosome in a typical instance may thus be defined as the point of origin of the central fibrils. It is usually at or near (often slightly below) this level that the flagellum acquires its limiting membrane as an extension of the surrounding cell membrane. A thickened ring in or just beneath the membrane, or fibrous connections between it and the kinetosome, commonly mark this point of emergence of the flagellum from the cytoplasm proper. This relationship holds even where the kinetosome is retracted more or less deeply into the cell, as in trypanosome flagellates and many other organisms. In such cases, the cell membrane is deeply invaginated; at the bottom of the invagination the membrane recurves to form the surface membrane of the flagellum, and the central fibrils make their appearance. It would seem that the presence of the central fibrils and of a flagellar membrane are almost always coincident in motile flagella.

Two sorts of exceptions to this rule of association have been reported. One is the cilia of *Euplotes patella,* investigated by Roth (1956). The core of the kinetosome in these cilia is occupied by tubular fibrils of about the same diameter and appearance as the central fibrils, but more sinuous. Whether these are direct continuations of the central fibers is not certain; a granule occurs at the apex of the kinetosome, and none of Roth's micrographs shows unequivocally that the fibrils traverse the granule.

Another intriguing exception is the spermatozoid of the brown alga, *Dictyota*, studied by Manton (1959b). When this cell is first liberated from the parent plant, it is coiled, and the whole length of the single functional flagellum is wrapped about the cell body, *under* the cell membrane. Sections show clearly that the 11 fibrils of the axis are all present, without a limiting membrane of their own. The kinetosome is recognizable by its lack of central fibrils and by the septum across its upper end, where central fibers arise. Soon after liberation, the flagellum lifts from the cell surface, the membrane closing around it and over the cell below it. This emergence appears to occur progressively from the distal tip toward the base, so that intermediate stages can be recognized in which the basal part of the flagellum lies within a slight ridge on the cell surface. It would be interesting to learn when the first movements of the flagellum can be detected in this cell.

A very similar observation, again by Manton (1959a), concerns the new flagellum in dividing cells of *Micromonas pusilla*. In a set of serial sections, the kinetosome and a short length of flagellum are clearly seen beneath the cell surface; only the two central fibers project as a minute membrane-covered finger. Emergence of the flagellum must involve a movement of the whole structure toward the surface, or a shrinking back of the cytoplasm around it.

Most interestingly, examples are known of non-motile cilia, associated with sensory structures in metazoa, in which no central fibers appear at all. This is true of the cilia connecting the inner and outer segments of the rod and cone cells of the vertebrate eye (DeRobertis, 1956), including the third or pineal eye of reptiles (Eakin and Westfall, 1959). The kinetosome here is distinguishable as a more compact proximal segment of the fibrous cylinder. During rod cell development (DeRobertis, 1956; Lasansky and DeRobertis, 1960; Tokuyasu and Yamada, 1959), the slender cilium grows out from the primordium of the inner segment, quite like a free cilium, except that central fibers are not present; the bulk of the cilium rapidly expands to form the distal segment of the rod. On the other hand, in another instance of a receptor-associated flagellum, the invaginated second flagellum of *Chromulina psammobia* (Rouillet and Fauré-Fremiet, 1958a), the flagellum is complete with central fibrils.

That animal centrioles are constructed on a plan closely similar to that of kinetosomes, as had long been suspected by light microscopists, was first shown to be true by de Harven and Bernhard (1956) for a number of kinds of vertebrate cells and has subsequently been demonstrated in innumerable animal species, invertebrate as well as vertebrate. Nine skewed triplet fibrils show very clearly in a centriole in a chick spleen cell pictured by Bernhard and de Harven (1960) and interconnections between subfibrils of adjacent fibrils, as well as a central cartwheel structure, appear distinctly in micrographs of snail sperm centrioles by Gall (1961). Animal cell centrioles typically are present at least in duplicate during interphase; each is a short, open, fibrous cylinder and the two generally are oriented at about right angles to each other. Sotelo and Trujillo-Cenoz (1958a, 1958b) have studied the origin of flagella from preexisting centrioles in developing spermatids of arthropods and vertebrates and in embryonic chick neural epithelium. Both in the spermatids and in the epithelial cells the centrioles migrate to the cell periphery, where one of them becomes oriented perpendicular to the surface and functions as a kinetosome. The peripheral flagellar fibrils either grow out from the kinetosomal fibrils or are organized from vesicles in the flagellar bud and become continuous with the kinetosomal fibrils. Central fibrils appear at the same time. Following establishment of contact or at least contiguity between the distal end of the flagellum-forming centriole and the cell membrane, the kinetosome withdraws into the cell, carrying with it the cell membrane wrapped around the young 9+2-patterned flagellum and lining the pit that it occupies. Subsequently, the pit may disappear as the kinetosome returns to the cell periphery.

No similar observations on the sequence of development of protozoan cilia are yet available, nor do we know how the fibrous structure of kinetosomes is reproduced when these organelles duplicate. Roth (1960b), studying division stages in the hypotrich ciliate *Stylonychia,* found groups of ciliary fibrils embedded in the cytoplasm, not immediately adjacent to existing intact cilia, at a time when new ciliary organelles are expected to be developing. Some of these appear to have central fibers, some do not, and some seem to be incomplete cylinders. The implication that kinetosomes are built up gradually from morphologically undifferentiated precursors is suggested, but, as Roth recognized, further study would be necessary to substantiate this. This is especially so since old ciliary organelles in the hypotrich ciliates often dedifferentiate while the new ones form or soon thereafter. Dedifferentiating flagella have been seen in the multiflagellated, syncytial zoospores of the alga *Vaucheria,* by Greenwood (1959). Upon settling on a substrate these organisms withdraw their flagella into the protoplasm. Sections show clearly that the 11 fibrils of the flagella are present in the cortex of the syncytium although the flagellar membranes have disappeared and no flagella are evident externally. Ehret and Powers (1959) argue that kinetosomes of gullet organelles of *Paramecium* arise *de novo*, physically separated from any preexisting ones, but their evidence on this score is derived from light microscope studies, which have led other workers to opposite conclusions, so the question must be regarded as unsettled.

For metazoan cells, the problem of centriole replication has been considered recently by Bernhard and de Harven (1960) and by Gall (1961). According to the former authors, scrutiny of centrioles in a variety of tissues often reveals the presence of "satellite" bodies, small dense structures about 70 mμ in diameter. Several of these bodies may appear anywhere within a zone of about 250 mμ or so around the cylindrical centrioles, and often some of them are united by slender dense bridges to the centriole proper. A single fortunate section of a cell from a leukemic mouse liver showed the two centrioles lined up one behind the other. An arrangement of fibrous structures at their adjoining ends distinctly suggested two additional, very short cylinders oriented at right angles to the mature centrioles. Bernhard and de Harven propose that these are daughter centrioles formed by a process of lateral budding from the mothers, and that the satellite bodies represent some sort of precursors. Gall's (1961) important paper likewise illustrates the appearance of short "procentrioles" at right angles to the parent centriole. In atypical spermatocytes of the snail *Viviparus,* multiple procentrioles — as many as eight in a single section — may be seen, lined up radially about one end of a mature centriole (these ultimately all mature and produce flagella). The often observed arrangement of two centrioles at right angles to one another thus may be explained as the persistence of a relationship established when a new centriole is formed. Later, the two centrioles may separate; an alternative position commonly observed is in single file, with a common long axis. This was true of the "budding" mother centrioles in the figure by Bernhard and de Harven and often obtains when one centriole in metazoan tissues is in the process of forming a cilium. Indeed, Bernhard and de Harven observed centrioles in this position, from one of which an abortive cilium was projecting into a depression of the cell surface, in a cell where cilia, according to all previous knowledge, had no business to be at all.

The whole question of genetic continuity of centrioles and of kinetosomes is one of long standing (see Lwoff, 1950; Gall, 1961). The equivocal nature of morphological data, even at the ultra-microscopic level, is not surprising. If new kinetosomes are organized independent of old ones, then there may be nothing to be seen until the new kinetosome is there, no way of recognizing that it is about to be there. On the other hand, if an old kinetosome is required either as a physical parent or as a template for the construction of a new one nearby, morphological problems are not much less formidable. The sample available to electron microscopy is so minute, in both time and space, that a rapid process occurring in a very small area is excessively difficult to catch. In protozoa, kinetosome replication commonly precedes any other sign of morphogenetic activity, so that selection of favorable cells for study is not easy. Probably the best bets are cells regenerating their surface structure following surgical mutilation, or species with such prolonged but rigidly scheduled differentiation programs that the time of replication of specific, identifiable kinetosome groups can be accurately predicted. In any case, long sequences of sections, covering enough area to permit accurate topographic identification, will be necessary. However, in view of the intense investigation under way in laboratories all over the world, employing morphological and biochemical methods, early resolution of some of these problems is to be expected.

The doubled condition of centrioles in resting animal cells reminds one of the fact that the great majority, at least, of flagellated protozoa bears two or more flagella. Usually the two kinetosomes of a biflagellate are close together but oriented at an angle (not necessarily 90°) to one another, or if parallel they show opposite polarity. As will be seen in the succeeding chapters, several presumed uniflagellates have turned out to have second kinetosomes which either are blind or terminate in short flagella not detectable with the light microscope. The weight of present evidence would indicate that in a few very simple cells a true uniflagellate condition obtains; that is, that kinetosomes can exist singly. Such appears to be the case in three small green flagellates (Chapter 4) and in trypanosomes (Chapter 5). Double kinetosomes often are shown in figures of these species but are interpreted as early division stages. We might say that kineto-somes *prefer* to come in pairs, but that they presumably can function temporarily, in the non-dividing cell, as single units. It is noteworthy in this connection that Mazia, Harris, and Bibring (1960) have demonstrated that the mitotic centers of echinoderm eggs are normally always duplex, but that the two components can be forced experimentally to separate and function singly as mitotic poles.

The discovery of the 9/9+2 pattern surely is one of the most provocative that electron microscopy has yielded. The universal occurrence of an unvarying geometric configuration at such a relatively gross multi-macromolecular level is without parallel. Unlike some other exciting ultrastructural generalizations that coincide satisfactorily with models previously conceived by biophysicists or at least invite rational interpretation, this one remains to date an almost complete enigma.

The most primitive protists known, the Monera, lack both centrioles and true flagella (pecilokonts) as well as membrane-limited nuclei, mitochondria, Golgi elements, and intracellular fiber complexes. Development of such discrete organelles to carry out the multiple assignments of cell metabolism in a manner efficient enough to allow larger cell size and greater structural elaboration than are achieved at the moneran level undoubtedly occupied a long and hectic chapter in the ancient history of life. One can only guess when in this period the kinetosome/centriole evolved (see Chapter 4), and whether its primary function was flagellar locomotion or nuclear division. Certainly it could only have been perfected by a cell in which mechanisms for the transmission of information from one generation to the next had reached a high level of precision.

Grimstone (1959b) proposes that flagella may have evolved independently in several groups, sharing a common structure down to its minutest details because no other can fulfill the mechanical and developmental requirements of such an organelle. But whether it evolved once or several times, the 9/9+2 pattern would hardly have been perpetuated so faithfully without good and sufficient reason; it cannot conceivably have been an evolutionary accident. Some types of cells accomplish nuclear division without detectable centrioles, but apparently no eucellular organism has achieved flagellar locomotion, or anything like it, without 9+2 fibrils arising from a kinetosome, which usually also is the point of origin of other, intracellular fibers. This means that the 9+2 arrangement must be indispensable to the mechanics of flagellar locomotion — no reasonable approximation will do — and that the cylinder of nine fibrils is one of very few successful designs for a cytoplasmic center of morphogenetic and kinetic activity. Regardless of whether the first centriole was a kinetosome or vice versa, the fiber arrangement is significant to both functions.

It is usually supposed that flagellum movement results from the passage of waves of localized contraction along some or all of the peripheral fibers (Gray, 1955; Afzelius, 1959; Gibbons and Grimstone, 1960). Inoué (1959), however, has suggested that only the central fibers may be contractile since these are the ones missing from kinetosomes and from non-motile receptor cilia. The nature of such contraction is not understood and cannot be understood until we know a great deal about the chemical and physical structure of the fibrils. Analyses of isolated cilia of *Tetrahymena* (Child, 1959; Watson, Hopkins and Randall, 1961) and flagella of some sperm and phytoflagellates (Tibbs, 1957, 1958; Jones and Lewin, 1960) show that they are largely protein, with varying amounts of loosely bound nucleotide, carbohydrate and lipid. Adenosine triphosphatase (ATP-ase) activity is present, as indicated also by the cytochemical studies of sperm tails by Nelson (1959), while Hoffman-Berling (1955, 1958b) and Bishop and Hoffman-Berling (1959) report ATP-reactivation of extracted sperm models. Nelson's electron-microscope cytochemical studies showed ATP-ase and succinic dehydrogenase concentrated in the nine peripheral fibers throughout their length; the other studies did not permit differentiation of fiber and matrix composition. Using immunologic techniques, Finck and Holzer (1961) found that proteins similar to actin or myosin were absent in cilia and sperm tails from the chicken.

Some information on the composition of kinetosomes is

reported by Seaman (1960), who employed a detergent for progressive solubilization of alcohol-killed cells of *Tetrahymena*. He was able to isolate large populations of uniform tiny particles that appear to be kinetosomes, although confirmatory electron-microscope examination of sectioned pellets has not yet been made. Comparison of the presumed kinetosomes with whole-cell extracts showed a significantly higher proportion of protein in the former (49·6 per cent dry weight versus 16·4 per cent in whole-cell extracts), less lipid and carbohydrate, and, most significantly, a much higher DNA:RNA ratio (1·5 in kinetosomes, 0·1 in whole cells), although total nucleic acid content was similar in both. Measurements of enzymic reactions showed a considerably greater specific activity in the kinetosomes than in whole-cell extracts for glycolysis, oxidative phosphorylation, apyrase, succinic dehydrogenase, and fumarase. In preliminary cytochemical studies, Randall and Jackson (1958) and Randall (1959b) reported the demonstration of DNA and RNA in kinetosomes of *Stentor, Tetrahymena,* and two metazoan epithelia.

An understanding of the structural chemistry of the flagellum and kinetosome will constitute a great leap forward, but thus far such efforts have provided no clues to the reason for the 9/9+2 pattern. Beyond recognizing that the fiber number and arrangement represent a functional necessity for kinetosomes/centrioles as well as for flagella, we have not progressed appreciably in this direction since the early speculations of Manton (1952, 1956), Astbury, Beighton, and Weibull (1955), and Bradfield (1955) (see also Serra, 1960). Although the skewing of doublet and triplet fibrils, the attachment of additional fibrous structures to the kinctosome-centriole, and the one-way orientation of arms on the peripheral fibers all confer a fundamental asymmetry on these bodies, there is superimposed a bilateral symmetry in the flagellum itself, as Manton (1952, 1956) has pointed out. Careful study of the arrangement of spines and mastigonemes relative to the 11 axial fibrils in some phytoflagellate flagella has led her to conclude that the plane of symmetry passes between the two central fibrils and bisects one of the nine peripheral fibrils. Mastigoneme rows are disposed symmetrically on the two sides. Fawcett and Porter (1954) noted that, in the molluscan gill epithelium they were studying, the two central fibers appeared to have a consistent orientation in all cilia of a field, and that a line connecting the two fibers was perpendicular to the plane of beat of the cilia. More recently Afzelius (1961) found a consistent bilateral arrangement of accessory lamellae in the packed swimming-plate cilia of a ctenophore, and established that the direction of active beat of the cilia was toward the single fibril bisected by the medial plane.

This consistency in orientation may be a general rule, although the high-resolution pictures required to demonstrate it are not always available. But it seems to apply frequently to ciliates as well as to ciliated epithelia, and in these protozoa the cilia are capable of bending in any direction, and commonly do deviate from the beating plane on their recovery stroke — a fact too often ignored but clearly demonstrated by Párducz (1954). The orientation of the central fibers might still be associated with a preferred beating direction, a relationship that remains to be explored.

The fundamental asymmetry of the flagellum may perhaps be related to another observation on ciliary movement by Párducz (1954). Ciliate cilia may under certain circumstances cease their directed, locomotor beating and move instead in a simple, ineffectual conical path, which the author considers to be a primary movement related to the simplest form of beat in a flagellum. In all of the many ciliate species examined this is a counter-clockwise sweep when the viewer observes the ciliated surface from above. As noted earlier, Gibbons and Grimstone (1960) and Gibbons (1961) found that the arms on the peripheral flagellar fibrils point counterclockwise if the flagellum similarly is viewed from its distal end.

Finally, one aspect of the problem that has not received much recent attention is that of polarity of the whole flagellar apparatus and the polarity imposed by it on the cell. In flagellates there typically is a spatial association, intermittent with mitosis or continuous, between nucleus and flagellar basal bodies. Golgi elements, contractile vacuoles, and other organelles occupy fixed positions also near the basal bodies. In ciliates, individual ciliary rows, as well as the whole body, are strongly polar and asymmetric. In all these organisms, kinetosome duplication precedes the visible inception of mitosis. Sequence and synchrony in morphogenetic events require some sort of directional information transfer among the various organelles involved; the cell and its components have to recognize, among other things, which end is up. The problem clearly is not limited to cells with flagella, or even with centrioles, but it is significant to note that Gall (1961) has found some evidence of a consistent relationship between asymmetry and polarity in sperm centrioles.

Et Cetera

Although trichocysts are hardly a common cell organelle, they occur in many ciliates and flagellates. Like flagella, they attracted the attention of early electron microscopists because they were accessible to study by techniques other than sectioning. A review of their structure here will provide a brief introductory notion of the sorts of fibrous elements that protozoa are capable of elaborating and that constitute an important aspect of differentiation in many of them.

Trichocysts are minute organelles present in the peripheral cytoplasm; under the influence of many sorts of stimuli they "explode", discharging long slender filaments through the cell surface into the environment, or sometimes, abnormally, into the internal cytoplasm. At least two kinds and perhaps three kinds of true explosive trichocysts occur. The most familiar is the conspicuous, non-toxic rod expelled by *Paramecium* and a number of other holotrichous ciliates (Jakus, 1945; Jakus and Hall, 1946; Pease, 1947; Knoch and König, 1951; Beyersdorfer and Dragesco, 1952a, 1952b; Dragesco, 1952a; Krüger, Wohlfarth-Bottermann, and Pfefferkorn, 1952; Nemetschek, Hofmann, and Wohlfarth-Bottermann, 1953; Wohlfarth-Bottermann, 1953a; Potts, 1955; Sedar and Porter, 1955; Rouiller and Fauré-Fremiet, 1957a), and also found in a flagellate (Dragesco, 1952b). The function of these trichocysts is enigmatic, in spite of considerable study of their structure. Whether the dense fringe of discharged trichocysts produced by *Paramecium* provides a protective cover is a moot point. Some authors have suggested that they have adhesive properties serving for attachment, but this has not been shown in recent studies. In media of relatively high salt concentration, salts are present as granular deposits in the shafts of discharged trichocysts, and Wohlfarth-Bottermann (1953a) suggests that they serve an osmoregulatory function.

A second type of trichocyst is toxic, causing paralysis and some cytolysis upon contact; it is characteristic of predatory holotrichs of the Order Gymnostomatida (Dragesco, 1952c). A third (on the basis of electron-microscope studies) has been described only from a cryptomonad flagellate (Dragesco, 1951).

The undischarged non-toxic trichocyst of the *Paramecium* type consists of a homogeneous spindle- or carrot-shaped body of low density, situated with its long axis perpendicular to the pellicle surface. Upon discharge it becomes a long, uniform or tapering shaft with a conspicuous cross-banding of about 55 mμ, capable of fraying into striated longitudinal filaments. A few micrographs of high resolution show a two- or four-band striation within the major period, and a super-period encompassing four of the major periods often appears. General similarity to the structure of collagen fibrils (Hodge, 1960) is striking. Early studies of whole discharged trichocysts indicated that the striated structure was a hollow cylinder, since it appeared very flat in the dried specimens. However, subsequent studies of sectioned material suggest that the shaft is solid, with its characteristic periodicity observable throughout; apparently the extreme flattening of dried shafts reflects a high water content. Jakus's (1945) extensive investigations showed that the trichocyst material was a protein similar to collagen in structure and extensibility, in swelling properties at pH's far from its isoelectric range, and in resistance to tryptic digestion. But other reactions were not typical of pure collagen, and the 55-mμ periodicity of trichocysts is significantly smaller than the 64-mμ period of collagen.

Since unextruded trichocyst bodies clearly do not contain folded-up, preformed trichocyst shafts, the problem of discharge is particularly significant. The process occurs in milliseconds. Rouiller and Fauré-Fremiet (1957a) succeeded in obtaining thin sections of *Frontonia* showing trichocysts in the process of discharging internally. Earliest stages show, within the homogeneous matrix, islands of organization where periodicity of 12 to 15 mμ is observable. In later stages, the periphery of the shaft is very regularly striated at about 24 mμ, and increasing zones within the matrix show the finer periodicity. Parts of a single trichocyst reach the mature, 55-mμ period while other parts are still organizing. The process involves (1) the unmasking or creation of molecular groupings capable of binding heavy metals such as osmium or tungsten used in the preparation of electron-micro-scope specimens, (2) the regular ordering of these groupings according to a crystalline pattern, and (3) apparently a considerable volume increase most easily attributable to water uptake; discharged trichocysts may be six to ten times as long as resting ones, and probably not correspondingly more slender.

The tip of the trichocyst of *Paramecium* is preformed in the resting state and discharged without alteration. It is an elongate, tack-shaped structure that does not flatten during drying and has a striation period of 17 to 30 mμ. Species of *Paramecium* differ consistently in the shape and size of the tip. In *Frontonia* the tip is elliptical, unstriated, and of low density, and trichocysts of some other ciliates apparently lack specialized tips.

In a peculiar small phytoflagellate of uncertain (dinoflagellate or cryptomonad) affinities, *Oxyrrhis marina,* Dragesco found slender trichocyst bodies arranged peripherally. On discharge these appeared as discrete fibers with a pronounced period of 60 mμ, resembling thin versions of the *Paramecium* rod.

The toxic trichocysts of 24 species of gymnostomes studied by Dragesco all are remarkably similar in morphology. In the intact cell viewed with the light microscope they are found anywhere around the periphery but most abundantly about the mouth; they are visible as elongate, banana-shaped bodies. On discharge they are visible in electron micrographs as flattened, empty capsules and long, slender, unstriated shafts. Occasionally the tip of the shaft appears thicker, and the author suggests that in this type of trichocyst a preformed tubule within the resting capsule is everted during discharge, incomplete eversion resulting in the thicker terminal segment.

Explosive trichocysts in the cryptomonad flagellate, *Chilomonas paramecium,* come in two sizes; large ones grouped about the anterior pharyngeal invagination and smaller ones on the general body surface. The latter appear in electron micrographs of dried preparations to consist of two uniform fibers with an irregular, bulbous, apical swelling from which a short, tapering twig emerges at an angle. Pharyngeal trichocysts are similar but longer, with several axial fibers.

In addition to the true explosive trichocysts, many protozoa possess mucigenic bodies, small globular elements typically

E

arranged in rows under the cell surface. Under certain conditions these may discharge their contents as sticky, ductile filaments. It is generally agreed that this material is a mucoid or gelatinous substance normally contributing to the formation of amorphous cyst or other, possibly protective, external coatings. Presumably the so-called protrichocysts of some ciliates are analogous structures. The latter appear as spheroidal or pear-shaped bodies of low density in thin sections of tetrahymenid ciliates (Pitelka, 1961a; Grassé and Mugard, 1961). Following discharge, regenerating mucigenic bodies in *Ophryoglena* (Grassé and Mugard, 1961) are visible as long rods with dense contents; their density decreases as they mature. In a brief, unillustrated abstract, Dragesco (1952d) states that discharged filaments of mucigenic bodies in several flagellates and ciliates are very variable in size and show no definable structure in electron micrographs.

Structures that in the light microscope appear to be fibers occur in nearly limitless variety of form and arrangement in protozoa, particularly in ciliates and flagellates. These cannot be described without detailing the morphology of the species concerned; hence they will be dealt with in the next chapters and summarized in the concluding section of this book. It may be noted, however, that the highly ordered structure of the discharged trichocyst of the *Paramecium* type is not unique. Other protozoan fibers, although constructed of quite different units, likewise display paracrystalline patterns of impressive precision.

Nuclei

In general, electron microscopy of nuclear structure has been disappointing. The problem, at least in protozoan studies, is not that nothing of interest has been seen, but that a number of different things have been seen, difficult to reconcile in terms of a unifying story of nuclear structure and behavior. This of course has always been true in protozoan nuclear cytology, but in recent years critical light-microscope investigations have permitted most of the divergent patterns and processes to be interpreted in conventional karyological terms (see Grassé, 1952; Grell, 1956; Nanney and Rudzinska, 1960); protozoa may be highly unconventional in some details of their mitotic and meiotic cycles, but they accomplish the same results, with the same basic materials, and according to the same fundamental rules as do other eucells. This is hardly surprising; ancestral protists presumably made the rules that metazoa and metaphytes have been following ever since, and exploratory variants of these rules have persisted and been embellished during the long evolutionary histories of divergent protistan groups. As many inspired light microscopists have realized, study of protozoan karyological pecularities is not merely esoteric.

Most electron-microscope studies of protozoa have included at least incidental observations on nuclear structure. We can in a general way list some properties that appear to be common to most of them, and then discuss the more spectacular exceptions. Our most consistent information concerns the organization of the nuclear envelope.

Surrounding the nucleus of all eukaryotic cells is an envelope consisting, in every case rigorously examined, of two distinct membranes separated by a low-density zone. The overall thickness of the envelope generally is in excess of 20 mμ and may be twice that or more, since the light layer or space varies considerably in width (Fig. 16, Pl. IV). A particularly elegant example of nuclear membrane structure is that revealed in Grimstone's (1959c) study of *Trichonympha*. The nuclear envelope here is perhaps thicker than in many other cells but its structure is probably typical. The two membranes of the envelope are each about 7 mμ in thickness and the intervening space averages 23 mμ. Distributed fairly regularly over the envelope are pore-like structures, or annuli, whose rims are formed by a fusion of the outer and inner envelope layers. Frequently, a layer of moderate density occupies the opening of the pore and may constitute a septum or plug across it. In surface view, the rims of the pores appear quite dense, partly owing to the presence of ten to 12 small dark granules regularly disposed around them; these do not appear to represent fibrils or cylinders, but rather distinct particles lying on the outer surface of the membrane. About 80 pores per μ^2 occur in *Trichonympha;* their inner diameter is approximately 45 mμ.

In many instances among multicellular organisms, and in some protozoa (Sager and Palade, 1957) continuities may be seen between the outer nuclear membrane and cytoplasmic membranes

(Fig. 16, Pl. IV), the peri-nuclear space thus being in communication with cavities of the endoplasmic reticulum. In some species of the flagellate *Chrysochromulina,* the outer nuclear membrane is continuous with the outer chloroplast membrane where the pyrenoid abuts against the nucleus (Manton and Leedale, 1961). Whether these continuities reflect a persistent functional relationship or not we cannot say. It is clear that in many metazoan cells the nuclear envelope forms during late mitosis from a halo of membranous reticulum (e.g., Barer, Joseph and Meek, 1959; Harris, 1961). The conclusion that cytoplasmic membranes, particularly the RNP-rich granular membranes, may originate at the nuclear surface has been tempting; however, to date no evidence of a clear gradient in concentration of Palade particles or membranes from the nucleus toward the cell periphery has been presented, even in cells in which active regeneration of ergastoplasm is known to be occurring (Bernhard, 1959).

Grimstone (1959c) has considered his micrographs of *Trichonympha* in the light of these data. He finds in the cytoplasm many elongate, flat membranous sacs bearing granules on their outer surfaces; no reticular interconnections appear. These are particularly abundant around the nucleus and may form a discontinuous shell over its surface. However, no direct continuities between nuclear and cytoplasmic membranes are seen, and the latter are considerably thinner than the former and also lack their annular structure. Grimstone illustrates clusters of unattached granules very close to the nuclear membrane, and he considers it likely that the membranes may be formed in the near vicinity of, but outside, the nucleus, and that the granules then are acquired by migration from the nuclear surface.

The majority of protozoan nuclei studied probably have envelopes similar to that of *Trichonympha,* although perhaps the annuli are less abundant (Fig. 17, Pl. V). Opinion differs as to whether the pores are patent. A striking elaboration of the nuclear envelope was found in *Amoeba proteus* first by Harris and James (1952) and Bairati and Lehmann (1952), and subsequently has been studied by other workers (see citations on amebae in Chapter 3). A porous double membrane exists as in other nuclei; immediately beneath this is a layer, up to 300 mμ deep, consisting of an orderly honeycomb arrangement of packed tubes (Figs. 18

and 19, Pl. V). These are open at their inner ends, and are roughly hexagonal in cross section, with a diameter of about 140 mμ; their walls appear dense in electron micrographs and may be double membranes. Where the tubes terminate at the nuclear membrane, each is concentric around a pore. Lehmann (1958b) noted the attachment of peripheral nucleoli to the inner ends of the tubes by strands of delicate vesicles. Similar material appeared within the tubes and outside the nucleus, and he speculated on the possible passage of vesicular material from nucleoli to cytoplasm in this manner. Mercer (1959) also observed vesicular material near the nuclear surface and postulated that it might originate from the nuclear membrane. However, such vesicular material is so similar to the wispy substance representing components of the cytoplasmic matrix after fixation that it would at best be difficult to demonstrate such a migration.

The nucleus of *Amoeba proteus* is large, up to 50 μ in diameter, and maintains its biconcave disc shape while floating in the streaming endoplasm. Consequently it has been suggested that the honeycomb layer of the nuclear cortex may have a function in the maintenance of form. It is extremely interesting to note, then, that more or less similar honeycomb layers have more recently been found in three other protozoan nuclei.

Beams, Tahmisian, Devine, and Anderson (1957) first reported on the nuclear envelope of *Gregarina rigida,* a sporozoan parasitic in the intestine of a grasshopper, and subsequently Grassé and Théodoridès (1957, 1959) published observations on a gregarine from a tenebrionid beetle. The honeycomb layer, only about 90 mμ deep in these organisms, has the appearance of a dense meshwork of filaments closely applied against the usual double nuclear envelope but regularly interrupted by bulbous or conical channels open to the nucleoplasm. Peripherally, the channels again appear to be concentric to pores in the outer envelope. The French authors, however, consider that the pore-like structures are vesicles in a continuous membrane.

In 1959, Beams and his colleagues found that *Endamoeba blattae,* (distantly related to *Amoeba),* occurring as a parasite in the hindgut of a cockroach, has an *Amoeba*-like honeycomb layer at the nuclear surface, but that in this case it occurs on the *cytoplasmic* side of the

PLATE V

FIG. 17. Tangential section at surface of nucleus of *Bodo saltans* showing pores or annuli, each with single dense particle at center. Abundant Palade particles on cytoplasmic side of nuclear envelope. × 45,000.

FIG. 18. Section through cortex of *Amoeba proteus* nucleus, showing double nuclear envelope and honeycomb layer inside envelope; irregular dense peripheral nucleoli.   × 27,000.   From Roth, Obetz and Daniels, 1960.

FIG. 19. Oblique section near surface of nucleus of *Amoeba proteus,* showing cortical honeycomb layer.   × 31,000.   From Roth, Obetz and Daniels, 1960.

FIG. 20. Cross-section through small part of metaphase spindle of *Amoeba proteus.* Tubular profiles of spindle fibers aggregated in clusters or bands, surrounded by clouds of fine material.   × 70,000. From L. E. Roth.

FIG. 21. Section from *Amoeba proteus* interphase nucleus showing helices linked in linear aggregates.   × 31,000.   From Roth, Obetz and Daniels, 1960.

FIG. 22. Small area from longitudinal section of *Amoeba proteus* metaphase nucleus, showing tubular spindle fibrils and small, densely fibro-granular chromosomes.   × 54,000.   From L. E. Roth.

The micrographs presented in Figs. 20, 22 and 23 have been published by L. E. Roth and E. W. Daniels, 1962, *J. Cell Biol.,* **12,** 57–78.

PLATE V

17

18

19

20

22

21

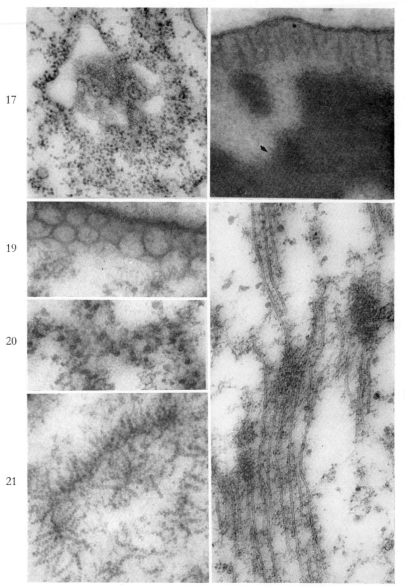

double nuclear envelope. The similarity of structure is unquestionable; the tubes in *Endamoeba* are somewhat more slender and perhaps not so regularly packed as those of *Amoeba*, but like the latter their walls appear more membranous than fibrous, and the layer is about 450 mμ deep. The same concentric arrangement of tubes about the pores of the nuclear envelope obtains in *Endamoeba*, but the tubes lead outward and open into the cytoplasmic matrix.

No figures are given for the dimensions of the gregarine nuclei, but it may be presumed that they are fairly large. Nuclei in the feeding gregarines are known to undergo considerable volume increase as the cell grows; measurements from a phase-contrast micrograph in a later paper of the Beams group (1959a) indicate a nuclear diameter of at least 20 μ. In *Endamoeba blattae*, the nucleus is up to 20 μ in diameter. Other than their size, which certainly is not unique, it is difficult to imagine what peculiar property in these organisms could be associated with the presence of this ordered structure. Electron micrographs of other gregarines do not show details of nuclear membrane structure, but certainly in *Pelomyxa* and several small lobose amebae, no honeycomb layer is present. The most striking fact emerging from these studies is this: whatever the function of the honeycomb layer, and whether or not it is homologous in the gregarines and the amebae, its perforations coincide with annuli in the double nuclear envelope. This indicates that the annuli have to be accessible to both nucleoplasm and cytoplasm and suggests that they have particular importance in some aspect of nucleo-cytoplasmic communication.

Since protozoan nuclear contents vary widely in their appearance in life or when stained for light microscopy, it is not surprising that generalizations on electron microscope structure must be so broad as to be practically meaningless. An "average" interphase nucleus contains a matrix of finely granular or filamentous material irregularly distributed throughout, with amorphous condensations often appearing peripherally or internally. Some of these condensations represent true chromatin bodies while others may be coagulation artifacts. Bodies that usually are larger, more granular, and more compact often are identifiable as nucleoli or as endosomes (Feulgen-negative bodies that probably are analogous to nucleoli but that typically maintain their identity through mitosis).

In *Amoeba proteus,* dense nucleoli are numerous, usually appearing peripherally, close to the honeycomb cortex. Within the diffuse filamentous matrix of the resting nucleus, strikingly well-defined helical fibrils have been demonstrated by Pappas (1956a, 1959), Mercer (1959), Manni (cited in Lehmann, 1958b), and Roth, Obetz, and Daniels (1960). These appear usually in stellate clusters, suggesting that individual filaments radiate from a common point of attachment, or as strings of clusters as in Fig. 21, Pl. V. Each individual helix is about 250 mμ in length and includes up to nine or so coils with a diameter of 30 mμ, the fibril itself being 8 to 9 mμ thick.

Roth, Obetz, and Daniels (1960) and Roth (unpublished) have examined a series of stages in mitosis in *A. proteus,* extending earlier observations of Cohen (1957). In prophase, the nucleus assumes a spherical shape, nucleoli migrate medially and subsequently disappear, and the honeycomb layer of the nuclear cortex vanishes. No helices are to be seen now within the nucleus, but an irregular, heavy network of dense material assumed to be chromatin appears. At metaphase, sizable gaps are apparent in the nuclear envelope. Tiny dense chromosomes, 0·5 to 0·8 μ long are arranged on an equatorial plate, and spindle fibers have appeared. These have the appearance of slightly sinuous tubules, about 15 mμ in diameter, and are arranged in irregular strings or clusters embedded in an amorphous material (Figs. 20 and 22, Pl. V). In anaphase, spindle fibers similarly are visible both between the separating chromosome groups and extending from them toward the ends of the spindle (Fig. 23, Pl. VI). In late anaphase, the nucleus is a very thin, cup-shaped body bounded by the double envelope which now is complete. The chromatin is less condensed than in metaphase. In telophase, chromatin occurs as a diffuse network; nucleoli are reappearing but the honeycomb cortex still is absent. Following cell division, a reconstruction phase occurs during which the honeycomb layer reappears, coincident with a remarkable elaboration of what seems to be nucleolar material (Fig. 24, Pl. VI). Extensive, somewhat convoluted sheets of this material connect corpuscular nucleoli in a wide peripheral zone of the nucleus; only occasional bits of less dense material identified as chromatin still are visible associated with nucleoli. At the same time, helices are reappearing, also located peripherally, close to the nucleolar material.

According to the account of these authors, the helices disappear as mitotic chromatin appears, and vice versa. No evidence was found to suggest that the helices were contributing through the compaction of their substance or in any other direct way to the formation of the early chromatin network. The visible morphology of the helices is so different from anything observed in any other cell that the only clues to their significance are those provided by wishful thinking.

A very different and equally intriguing appearance is presented by the giant chromosomes of some dinoflagellates. These persist throughout interphase as long, coiled organelles attached to the nuclear membrane by their chromomeres. Grassé and Dragesco (1957) have examined species of *Gymnodinium, Amphidinium, Polykrikos,* and a symbiotic zooxanthella from a sea anemone, and Grell and Wohlfarth-Bottermann (1957) have pictured *Amphidinium elegans*. In all of these species, interphase chromosomes in thin sections (Fig. 25, Pl. VI) are seen to contain many filaments, each 3 to 8 mμ (Grell and Wohlfarth-Bottermann) or 13 to 14 mμ (Grassé and Dragesco) in diameter, coiled together in a helical bundle, the coils being separated by a relatively coarsely granular material. The French authors believed that the bundle represents the chromonema and the individual filaments nucleoprotein complexes. The coiling of the bundle is much grosser than that of the filaments in the *Amoeba* nucleus, but the individual filaments may be roughly of the same order of thickness. Grassé and Dragesco reported occasional observations of the zooxanthella chromosomes suggesting a finer coiling of filaments within the major helical bundle. They believed that the granular nature of the inter-fibrillar material probably was the result of coagulation during fixation of a matrix substance.

The macronuclei of ciliate protozoa generally appear by light microscopy to be compact, as compared with the vesicular nuclei of other groups. In electron micrographs a porous, double nuclear envelope is seen to enclose a fine fibro-granular reticulum set with randomly distributed, dense, fibrous or granular bodies of irregular shape. These in a ciliate such as *Paramecium* (Fig. 27, Pl. VII) (Tsujita, Watanabe, and Tsuda, 1957; Ehret and Powers,

PLATE VI

FIG. 23. Longitudinal section through one end of *Amoeba proteus* anaphase nucleus, showing nuclear envelope fragments around chromosomes and spindle. × 17,000. From L. E. Roth, unpublished.

FIG. 24. Section through part of *Amoeba proteus* nucleus in early post-mitotic reconstruction. Nucleolar material (N) present as granules and ramifying sheets. Honeycomb layer of cortex beginning to reappear. × 3,200. From Roth, Obetz and Daniels, 1960.

FIG. 25. Section of single interphase chromosome of the dino-flagellate, *Gyrodinium* sp. × 65,000. From Grassé and Dragesco, 1957.

PLATE VI

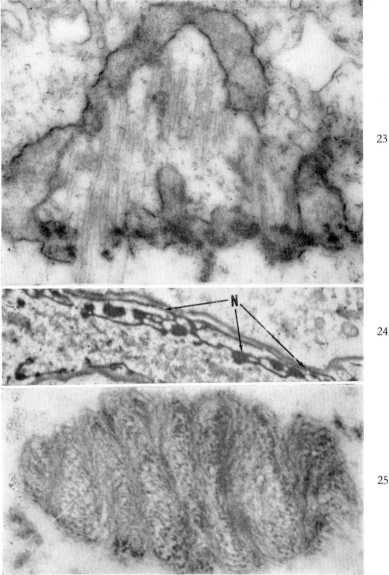

23

24

25

PLATE VII

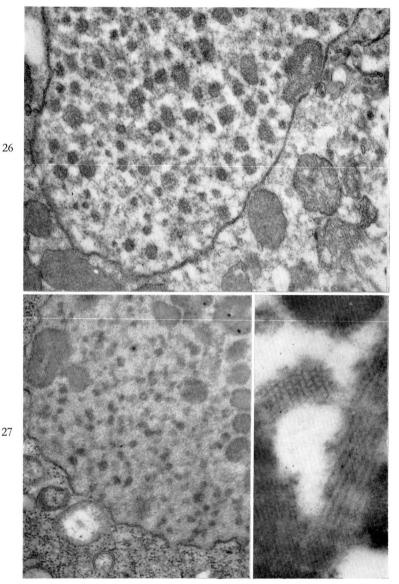

26

27

28

PLATE VII

FIG. 26. Part of normal macronucleus of *Colpidium campylum,* showing double envelope, presumed chromatin bodies and one nucleolus at right. × 22,500.

FIG. 27. Part of normal macronucleus of *Paramecium aurelia,* showing double envelope, numerous small chromatin bodies, and several larger nucleoli. Note abundant Palade particles in cytoplasm. × 21,700. From R. V. Dippell, unpublished (micrograph by K. R. Porter).

FIG. 28. Section through part of macronucleus of the suctorian ciliate, *Tokophrya infusionum,* from an overfed individual. Crystalline arrangement of chromatin material seen in longitudinal (parallel lines) and cross (honeycomb) section; normal chromatin body, at top, is dense and spongy. × 75,000. From Rudzinska, 1956.

1955) or a tetrahymenid (Fig. 26, Pl. VII) average about 100 to 200 mμ in diameter, and they are believed to represent the Feulgen-positive material of the macronucleus. Larger, more discrete dense bodies are probably nucleoli.

The ciliate macronucleus is hyperpolyploid, a small fragment sufficing to regenerate a genetically normal whole (see Sonneborn, 1949; Fauré-Fremiet, 1953; Grell, 1956). As is well known, it is derived from division products of the micronuclear synkaryon following sexual reproduction. Cytological and cytochemical methods show a many-fold increase (up to 200 ×) in DNA content in the macronuclear anlage, associated with repeated endomitoses; ultimately the macronucleus is totally responsible for nuclear-controlled phenotypic characters of the cell.

The indefinite survival of amicronucleate races of ciliates proves that in them a periodic renewal of the macronucleus is not essential. Many ciliate macronuclei, however, in addition to being replaced following sexual processes, are periodically reorganized in some manner, as evidenced by a condensation and change in staining properties accompanying amitotic vegetative division, or by an interphase phenomenon of progression of visible reorganization bands through elongate macronuclei. The significance of these processes relative to the necessary doubling of the macronucleus has until recently been unknown. Happily, a critical electron microscope examination of the reorganization phenomenon in the hypotrich, *Euplotes eurystomus,* by Fauré-Fremiet, Rouiller, and Gauchery (1957) was followed by a cytochemical and tracer study of the same species by Gall (1959 — Gall, however, was unaware of the earlier work).

In *E. eurystomus,* the very long, slender macronucleus in early interphase is filled with small, basophilic, Feulgen-positive granules. At a time about halfway between fissions, clear, weakly staining, transverse bands appear at both ends and progress slowly toward the center, where they disappear a few hours before the next division, leaving in their wake coarser granules showing more intense basophilia than before. In electron micrographs the reorganization band is easily recognized; at its medial (advancing) face, dense 1 μ bodies corresponding to those seen in the light microscope are abruptly cut off, apparently by a process of disaggregation, and are replaced by a fine reticulum in the reorganization band. High-magnification pictures show that the dense bodies are composed of very fine (about 8 mμ) convoluted filaments in compact masses. The reticulum of the reorganization band is likewise composed of interlacing, possibly spiraled fine filaments. Distal to the reorganization band, dense bodies reappear, showing a graded increase in size over a space of less than a micron. These do not differ appreciably from the pre-existing granules. Thus the reorganization appears to be a process of dispersion and reaggregation of a finely filamentous material making up the DNA-containing macronuclear granules. That the same phenomena characterize reorganization in *E. patella* is demonstrated by the observations of Roth (1957).

Gall's autoradiographic and photometric observations show that DNA synthesis (measured as incorporation of tritiated thymidine) occurs in a zone at the distal or posterior end of the reorganization band, and that the total quantity of DNA in the macronucleus is doubled by the time the bands meet. A basic protein, presumably a histone, increases in a similar manner, although precise quantitative data are not available. Gall's evidence thus suggests "that the micronucleus contains a number of DNA-histone 'units', presumably chromosomes, each of which duplicates once and only once" (p. 295) in each interfission period. To the general observer, the most interesting fact emerging from a comparison of the papers by the electron microscopists and by Gall may be that the chromosomal material is duplicated only in a disaggregated state.

Normal amitotic division of the macronucleus in *Stylonychia* (Roth, 1960b) and *Tetrahymena* (Roth and Minick, 1961) is accompanied by a coarse aggregation of granular elements and by the appearance of fine (30 mμ) fibrils in the constriction zone.

A unique nuclear structure has been reported by de Puytorac (1959b) in the macronucleus of the astome ciliate *Haptophrya plethodonis*. Here a dense chromatin network encloses many round, vesicular nucleoli, each surrounded by a weakly-staining aureole. In the electron microscope the latter is seen to contain concentrically arranged, discontinuous lamellae, suggesting (but high-resolution studies are needed) an intranuclear membrane system not found elsewhere.

Another curious nuclear inclusion was found in the macronuclei of two mating types of a strain of *Paramecium caudatum* by Vivier and André (1961). In addition to the usual dense, granular chromatin and nucleoli, they illustrated bundles of fibrils coursing randomly through the nucleoplasm. The fibrils were dense, about 25 mμ in diameter, and occasionally showed local zones of hexagonal packing. Larger bundles were detectable with light optics, and cytochemical tests showed that they contained neither DNA nor RNA. They were negative to Sudan and periodic-acid Schiff stains, but strongly positive to several protein tests. They were observed over a period of several years in healthy, sexually potent cultures, but possibly were more abundant in old cultures than in young ones. Thus there was no evidence that they accompanied any pathologic condition, and their significance is a mystery.

An extraordinary instance of highly organized structure in the DNA-containing material of the ciliate macronucleus was encountered by Rudzinska and Porter (1955) in the suctorian ciliate, *Tokophrya infusionum*. Vegetative reproduction in these suctoria is accomplished by endogenous budding, and the parent cell maintains its identity through its reproductive period, then ages and dies — a condition of cellular mortality not common elsewhere among the protozoa. The macronucleus divides very unequally during reproduction, the greater part of it being retained by the parent.

In an adult cell the macronucleus contains many small (0·5 μ), Feulgen-positive bodies, and in electron micrographs these appear as very dense, discrete spherules, composed of a compact filamentous meshwork like the chromatin bodies of many other interphase cells. In older cells, certain macronuclear granules appear hollow by light microscopy, and additional, irregular, wispy, Feulgen-positive bodies are seen. In electron micrographs the hollow granules are seen to be composed of material with the same density as before but showing a regular crystal-like order (Fig. 28, Pl. VII). The pictures suggest a close hexagonal packing of long, fine cylinders. The wispy material is composed of small clusters or strands of the same organized structure. Older cells of *Tokophrya* commonly undergo hemixis, a repeated division of the macronucleus unaccompanied by cell division, and Rudzinska and Porter speculate on the possibility that the appearance of the wispy material represents an evolution of chromatin bodies by condensation of nucleoprotein which aggregates then to form hollow granules and ultimately the patternless normal chromatin granules. However, no similar process was noted in normal young reproducing cells. Interestingly enough, Rudzinska (1956, and references cited there) found that overfeeding of young animals greatly reduces their life span and leads to hemixis and other changes associated with age. Their macronuclei show large chromatin masses with the same ordered structure as those of old cells.

Similar structural organization in the chromatin of developing spermatid nuclei of invertebrates has been demonstrated by a number of investigators (Yasuzumi and Ishida, 1957; Fawcett, 1958). Ultimately in these cases the crystalline pattern disappears as the nuclear contents become so dense as to conceal any internal structure. Presumably, changes in morphology from moderately dense and homogeneous to lamellar to crystalline to highly condensed coincide with alterations in the nuclear proteins and their bonding with DNA. At any rate, no degenerative change in the chromatin can be involved.

Morphological evidence of the movement of high-molecular-weight substances from nucleus to cytoplasm has been sought by many electron microscopists; several examples of pictures that could be interpreted as illustrating such a passage — through pores or via membranes — have been cited above. Additional examples susceptible to such interpretation are available in literature on ultrastructure of metazoan cells (Bernhard, 1959). But in no case are unequivocal sequences of stages shown. The paucity of evidence is rather surprising in view of the frequency of observations on nuclear ejections in the light-microscope literature. Ehret and Powers (1955) present phase micrographs of unfixed squash preparations of *Paramecium bursaria* indicating that intact young nucleoli are extruded from developing macronuclear anlagen following conjugation, but they could not illustrate this process in electron micrographs. However, they noted a general similarity in appearance between young nucleoli and small dense mitochondria in the perinuclear cytoplasm.

For *Amoeba,* an abundance of data is available from biochemical and tracer studies (see Brachet, 1958, 1959; Plaut, 1959). For example, it is concluded that RNA is synthesized in both nucleus and cytoplasm and that some movement of RNA or its immediate precursors from nucleus to cytoplasm occurs. But a large gap still exists between these data and the morphological evidence thus far assembled. It is worth while to note and collect suggestive morphological evidence; some of it eventually should turn out to be significant. But it is decidedly premature to say that any of it is more than suggestive.

CHAPTER 3

RHIZOPODS, ACTINOPODS, SLIME MOLDS, SPOROZOA

A PROPER phylogenetic approach to a discussion of protozoan morphology would require that we consider first the flagellates, but for the purpose of this review we shall abandon phylogenetic propriety and begin with the amebae. Although they are not primitive, their specializations are not primarily morphological — at least not at the level yet visualized by electron microscopy. They have for the most part discarded in the adult cell that remarkable organelle, the flagellum, and their most conspicuous common characteristic is the absence of an architecturally elaborate cortex and pellicle. Hence they may serve us, as they have served generations of text-book writers, as a convenient introduction to the cell types we will consider. Indeed, many of their characteristics have already been described as examples of protoplasmic structures in the preceding chapter.

The rhizopods, including the naked and testate amebae and the foraminifers, are placed by Grassé (whose taxonomic scheme, as presented in his *Traité de Zoologie,* 1952–53, is followed here) in a Subphylum Rhizoflagellata with the flagellates, from which they probably originated polyphyletically. Amebo-flagellates, showing characteristics of both groups, unfortunately have not been examined in detail by electron microscopists. The radiolarians, acantharians, and heliozoans are given separate status by Grassé as a Subphylum Actinopoda; many of them also have flagellated gametes. Actinopods have in common the possession of very slender pseudopodia, called axopodia, radiating from a generally spherical body. These differ conspicuously from the lobose pseudopodia of many amebae, but in some respects resemble the reticular fine pseudopodia of the Foraminifera or the filopodia of some other ameboid organisms. Jahn, Bovee, and Small (1960)

71

have recently suggested a radical revision of the taxonomy of these groups based on the kind of pseudopodia and the nature of their movement.

Superclass Rhizopoda

Amebo-flagellates, placed by Chatton (1953, in the Grassé treatise) as the first suborder of amebae, are remarkable for their double identity. Perfectly conventional small amebae, they are capable of transforming for short or prolonged periods into zooflagellates, reproduction occurring in either or both phases. Brief electron-microscope observation of *Tetramitus rostratus* (Pitelka and Schooley, 1955; Pitelka and Balamuth, unpublished) demonstrated that in the flagellate phase only, the organism has a pellicle reinforced with evenly spaced, parallel, fine fibrils, very similar to the pellicle of some simple zooflagellates (*e.g., Trypanosoma*, p.137). Whether the whole body surface is so clothed is not known. In fragmented cells, tracts of fibrils remain attached to kinetosomes of the four anterior flagella, suggesting a physical linkage.

A single electron micrograph published by Cigada and Cantone (1960) depicts a small monoflagellate cell believed to be a gamete of *Amoeba spumosa*. Unfortunately, not enough morphologic detail is included to permit comparison with other flagellates. Flagellate stages of rhizopods otherwise have not been examined.

The common, naked, lobose amebae have been for many years such attractive subjects for experimental studies in cell biology that it is not surprising to find a fairly impressive number of electron-microscope investigations devoted to them. By far the most popular forms have been the giants, *Amoeba proteus* and species of *Pelomyxa*.* Quite recently, several experimentalists and

---

* Since the publication in 1926 by Schaeffer of a taxonomic study equating the multinucleate *Pelomyxa carolinensis* Wilson with *Chaos chaos* Linnaeus, many biologists have adopted this usage. Chatton and Grassé (in Chatton, 1953) retain the genus *Pelomyxa* but follow Schaeffer in synonymizing *Amoeba proteus* Pallas with *Chaos diffluens* Müller. Other experts, among them Kudo (1959), conclude that *Chaos chaos* is unidentifiable from the early descriptions and that both *Amoeba* and *Pelomyxa* therefore are valid genera. The nomenclatural problem is even more chaotic than indicated here, and it seems expedient to follow Kudo in using the more prosaic names.

morphologists have turned to some of the abundant smaller species in related families, as being less spectacular but perhaps more representative both as rhizopods and as cells than their enormous cousins. Among the smaller forms, *Hyalodiscus simplex* (Wohlfarth-Bottermann, 1960b), *Hartmannella rhysodes* (Pappas, 1959), *Acanthamoeba* sp. (Mercer, 1959), and *Hartmannella astronyxis* (Deutsch and Swann, 1959), and the parasitic amebae, *Entamoeba invadens* (Barker and Deutsch, 1958; Deutsch and Zaman, 1959; Zaman, 1961), *E. histolytica* (Osada, 1959; Wohlfarth-Bottermann, 1959a), and *Endamoeba blattae* (Beams, Tahmisian, Devine, and Anderson, 1959b) have received at least passing attention. Chief sources of data on *Amoeba* and *Pelomyxa* are papers by Lehmann and his colleagues (1950, *et seq.*; Bairati and Lehmann, 1952, *et seq.*), Pappas (1956a, 1956b, 1959), Cohen (1957), Greider, Kostir, and Frajola (1958), Mercer (1959), Schneider and Wohlfarth-Bottermann (1959), and Roth (1960a).

The mobile surface membrane or plasmalemma consists of a unit membrane coated on the outer side (at least in *Amoeba, Pelomyxa,* and *Hyalodiscus* [Fig. 29, Pl. VIII], but not in some of the small amebae) with a diffuse, low-density layer of material bearing a continuous fringe or pile of very fine, hair-like extensions. These filaments, measuring 5 to 15 mμ in diameter and up to 150 mμ in length, are darkened by phosphotungstic-acid staining. Their dimensions suggest filamentous macromolecules, and their presence on the cell surface (where cytochemical tests show the occurrence of polysaccharide as well as protein) has prompted guesses that they are mucoproteins serving for adhesion to a substrate or to particulate food (Lehmann, Manni, and Bairati, 1956), and that they function in the surface binding of substances in solution shown to occur prior to pinocytosis (see discussion in Chapter 2). The hairy coating is very quickly lost from the surface of a newly formed food vacuole membrane (Roth, 1960a) but apparently persists at least for a limited time on deep pinocytosis channels and vesicles. It seems to be markedly reduced on the body surface of dividing *Amoeba* (Roth, Obetz, and Daniels, 1960); at this time the cell assumes a roughly spherical form, called the mulberry stage, with an all-over lumpy surface.

The cytoplasmic ground structure of amebae is of compelling

PLATE VIII

FIG. 29. Section through part of *Hyalodiscus simplex,* showing general features of cell structure in small, free-living ameba. Nucleus at lower left. × 18,700. From Wohlfarth-Bottermann, 1960b.

PLATE VIII

29

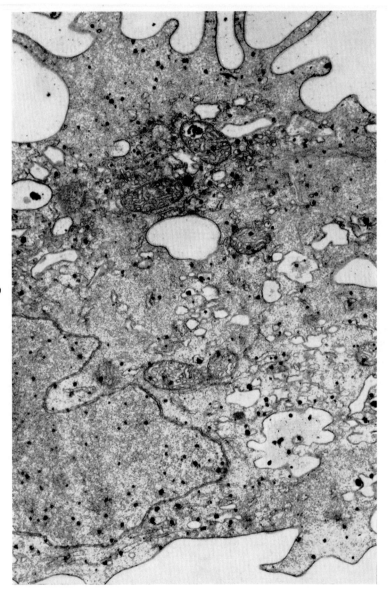

interest because of the perennial problem of ameboid locomotion and its importance in the general question of protoplasmic movement and sol-gel transformations. A digression to consider this problem seems in order. The predominant theory of recent years, in essential agreement with Mast's proposals of 35 years ago (see, for example, Noland, 1957; Landau, 1959) holds that forward flow of solated endoplasm results from contraction of plasmagel in the posterior end of the moving cell. Fluid proto-plasm reaching the tip of an advancing pseudopodium undergoes gelation at its periphery to form a plasmagel sleeve. Contraction of the gel in the posterior zone is followed immediately by solation, thus replenishing the internal plasmasol. According to Allen and co-workers (Allen and Roslansky, 1959; Allen, 1960, 1961; Allen, Cooledge, and Hall, 1960; Griffin and Allen, 1960), this theory is too simple. They find that the endoplasm is not uniformly solated and that hydrostatic pressure exerted by posterior gel contraction does not suffice to explain forward flow in such a non-Newtonian fluid. An axial core of endoplasm exhibits a viscosity nearly equal to that of the gelated ectoplasm. The authors postulate that streams or strands of cytoplasm move forward in this axial endoplasm to an anterior "fountain zone", where they pass outward to join and extend the ectoplasmic sleeve. Contraction at the anterior end, in the fountain zone, would actively pull the axial endoplasm forward. Observations on naked endoplasmic fragments confined in capillary tubes indicated that multiple individual streaming units could exist side by side.

The Allen group's hypothesis would seem to diminish the fundamental distinctions between pseudopodial movement of the lobose type and that of the filose type. Jahn and Rinaldi (1959) have reexamined the latter phenomenon as exemplified in *Allogromia laticollaris*. This genus is recognized by most authors, including Jahn and Rinaldi, as a foraminifer but is placed by Deflandre (1953, in the Grassé treatise) with the shelled amebae, some of which exhibit the same sort of movement. From the protoplasmic mass at the test opening of this organism a reticulum of fine, branching and anastomosing pseudopodia extends out over a radius of up to 15 mm. Because the pseudopodial surfaces are studded with adhering particles of food and other granules, streaming may readily be watched in the light micro-scope, and it always occurs in two directions simultaneously in every pseudopod. Granules pass outward along one side of a pseudopodium and, if it has a free end, turn around and pass inward along the other side. Multiple paths of streaming are visible on thicker basal pseudopodia and on protoplasmic nodes within the reticulum. Constant bending, twisting, lateral move-ment, splitting and joining of pseudopodia occur, but the two-way streaming is rarely if ever interrupted. Jahn and Rinaldi suggest that all but the finest pseudopodia are fascicles of finer units and that a fundamental streaming unit of probably less than 1 μ diameter exists.

The pseudopodia of *Allogromia* possess considerable rigidity, since they extend unsupported through the medium and do not collapse under the impact of collision with an errant ciliate. Granules in a single streaming path move all at the same velocity, like steps in an escalator. There is no evidence of any axial supporting structure nor of an internal, more fluid zone. Jahn and Rinaldi conclude that active shearing forces operate between paired hemicylindrical filaments in each pseudopod. The mechanism for such a continuous shifting of two gel surfaces against each other is unknown, but possibly would have some-thing in common with the phenomenon of sliding filaments believed to be responsible for the contraction of striated muscle. Similar processes may operate wherever cyclosis occurs in plant and animal cells, provided that the streaming mass has a relatively high viscosity so that a shearing force could act between its surface and a fixed cortical gel.

Both Allen's and Jahn's explanations of rhizopod movement require the assumption that streaming units possess considerable structural integrity. The stage would seem to be set for some critical investigations combining electron-microscope studies of ultrastructure with other biophysical techniques — if only the electron microscopists can discover some meaningful structure!

For the most part, results to date have been disappointing, as author after author has reported no detectable difference between solated and gelated parts of the cytoplasm of *Amoeba* and *Pelomyxa*. The most enlightening reports yet available are those by Wohlfarth-Bottermann (1960b, 1961) cited in Chapter 2, of studies on *Hyalodiscus simplex* (Fig. 29, Pl. VIII). In this relatively small ameba, viewed with the light microscope, the granular endoplasm occupies a spheroidal hump while the clear hyaloplasm surrounds it and forms a broad, flat anterior lobe or numerous small pseudopodia. Constant streaming of particles within the endoplasmic hump indicates that it is of fairly low viscosity. No sharp boundary exists between endoplasm and hyaloplasm; rather, a transition zone appears, wherein particles are displaced continuously or in little hernias towards the ectoplasm. In the hyaloplasm itself, the absence of inclusions prevents light-microscope discrimination of gelated and solated zones; however, the behavior of the exclusively hyaline pseudopodia proves that local transformations occur.

The light, nearly homogeneous texture of the cytoplasm described earlier is seen in Wohlfarth-Bottermann's electron micrographs around pinocytosis vesicles and in small pseudopodia where active flow presumably was occurring at the moment of fixation. Areas presumed to be gelated display a fine, compact, but not detectably orderly, net-like structure, and a continuous spectrum of intermediate conditions can be seen. Interestingly, many submicroscopically small pseudopodia are present on both dorsal and ventral surfaces of cells fixed *in situ* during locomotion on a glass surface, indicating that contact with the substrate occurs only at pseudopodial tips and that advance is stepwise rather than an overall rolling or flowing process.

The narrow transition zone between the hyaline layer and the endoplasm in *Hyalodiscus* shows merely a scattering of cytoplasmic particulates that become more abundant in the endoplasm. The ground substance of the latter is a loose granular reticulum resembling an intermediate condition seen in parts of the ectoplasm.

In earlier papers, Wohlfarth-Bottermann (1959a) and Schneider and Wohlfarth-Bottermann (1959) reported that in many different kinds of cells the gelated condition, or a state of high viscosity, is correlated with an increased compactness of the ground substance. Where active protoplasmic streaming is believed to be in process at the time of fixation, the cytoplasmic ground substance appears looser in texture. In some instances the compactness clearly accompanies general withdrawal of cytoplasmic water, as in the dark, dense, strongly dehydrated spore of the slime mold, *Didymium*. But in others, including a streaming channel of the slime mold plasmodium, textural differences are observed within very small regions of the same cell. In many instances, lamellar or vesicular elements and large cytoplasmic inclusions are absent from the zones assumed to be in a gel state.

Lehmann and various colleagues (see especially Lehmann, 1958a) have demonstrated repeatedly that the application of fixatives other than the customary, slightly alkaline, osmium tetroxide to *Amoeba proteus* can result in a distinctly reticular or fibrous structure in or just beneath the plasmalemma and around the wall of the contractile vacuole. They have not yet succeeded in compounding a fixative that preserves both this structure and that of other cytoplasmic organelles. Much of their evidence has come from isolated cell fragments treated with various acidic mixtures, and many workers consider that more artifacts may be introduced than avoided by these techniques.

That local variations in cytoplasmic texture can be detected at high resolution in well-fixed cells is conclusively demonstrated by Wohlfarth-Bottermann's work; this much is encouraging. The extremes of these variations are less spectacularly different than we might have hoped, and this means that correlating them with known physical states or activities will be no easy matter. They are even less conspicuous in the giant amebae that have been most frequently used for analyses of movements than in the comparatively neglected smaller forms. Furthermore, the existence of tiny pseudopodia and surface irregularities unsuspected by light microscopists suggests that underlying processes must be examined on a more minute scale than most workers have attempted.

Except in inclusion-free regions of the ectoplasm, the cytoplasm of *Hyalodiscus,* and other small amebae, contains membranous fine tubules and vesicles of varying sizes and shapes (Fig. 29, Pl. VIII). In *Amoeba* and *Pelomyxa* the cytoplasm is virtually filled with such vesicles. Variations in the appearance of the vesicle membranes and their possible significance have been discussed in the section on pinocytosis and phagocytosis in the preceding chapter. Most of the cytoplasmic membranes in the giant amebae are smooth, but occasional ones are seen bearing

Palade particles. The latter are sparsely distributed also in the cytoplasmic matrix — an observation that may be correlated with the generally diffuse basophilia of the large amebae. The smaller amebae tend to have more abundant particle-studded membranes and more basophilic cytoplasm.

Cytoplasmic inclusions generally recognized by light microscopists in the giant amebae and characterized by their stratification in centrifuged specimens as well as by their staining reactions include the following:

(1) Tiny, non-staining alpha granules, not mentioned by electron microscopists except Cohen (1957), who believed that they were represented by small, dense particles enclosed in membranes; their nature and significance remain unknown. Mercer (1959) saw similar small granules but did not attempt to identify them.

(2) Beta particles, larger than the alpha particles, now well identified as mitochondria. These are of the conventional microtubular type and show no special properties except for the peculiarly patterned mitochondria found in *Pelomyxa* during mitosis by Pappas and Brandt (1959) and described in the preceding chapter. The halo of mitochondria about the contractile vacuole has also been noted.

(3) Refractile bodies, unidentified in electron micrographs.

(4) Contractile vacuoles, discussed in Chapter 2.

(5) Plate-like or bipyramidal crystals in vacuoles. These are clearly recognizable in electron micrographs as rather large vacuoles of irregular shape within which a polygonal empty space appears. The crystals apparently survive fixation and embedding but either are torn out of the embedding plastic during sectioning or are quickly evaporated under the electron beam. Analysis of isolated crystals from both *Amoeba* and *Pelomyxa* has recently been reported by Grunbaum, Møller, and Thomas (1959) and by Griffin (1960), using chemical, physical, and optical methods. The crystals are identified as carbonyl diurea or a closely related compound; slight differences in composition or the presence of additional substances in small amounts may account for the different characteristic crystal shapes. The crystalline substance is believed to be a nitrogenous excretion.

Membrane-enclosed fat droplets are often seen in the cytoplasm.

In *Hartmannella* (Pappas, 1959) these are in intimate contact with mitochondria, an association often observed in metazoan cells. Although many light microscopists have been unable to identify Golgi elements in *Amoeba,* small piles of thin flat membranous sacs with adjoining microvesicles have been seen in *A. proteus* and most of the other amebae studied. By morphological criteria these clearly are classifiable as Golgi bodies.

Other particulates seen in electron micrographs but of unknown significance include very dense granules, 30 to 50 mμ in diameter, liberally distributed through cytoplasm and nucleus, occurring even within mitochondria, and sometimes concentrated on surface or vacuolar membranes.

An interesting recent observation by Roth and Daniels (1961) concerns infective organisms occurring in the cytoplasm of *Amoeba proteus.* In electron micrographs these were visible as single or multiple rod-shaped or spiral structures resembling bacteria, enclosed in vesicles. With phase contrast microscopy, the authors observed rupture of the ameba cell and its enclosed vesicles and the release of motile organisms into the surrounding fluid. Attempts to eliminate the infection by starvation, penicillin treatment and irradiation were unsuccessful. As the authors point out, these inclusions may be significant in view of the common use of *Amoeba* for studies of DNA and RNA synthesis. Evidence of cytoplasmic activity has been brought forward by workers who have suggested as one possibility the presence of symbiotic organisms.

Cytoplasmic structures of the parasitic amebae do not differ conspicuously from those of the free-living forms. However, in *Entamoeba invadens,* Deutsch and Zaman (1959) found no structures identifiable as mitochondria, and Osada (1959) reported the absence of mitochondria from *E. histolytica.* The same is true of some other parasitic protozoa, in which, presumably, energy release by electron transfer to oxygen does not occur.

Precystic individuals of species of *Entamoeba* typically produce crystalloid inclusions called chromatoid bodies, recognized by their strong refringence in life and by their pronounced basophilia. They gradually disappear in the cyst. According to Barker and Deutsch (1958) and Deutsch and Zaman (1959), precursors of the chromatoid bodies in feeding stages of *E. invadens* appear as small clusters of dense, 20-mμ particles, often near food vacuoles. These aggregate in large, rod-shaped masses showing a regular crystal-line order; the rods may reach 10 μ in length in the cyst. Cyto-chemical studies by these authors revealed a high RNA content, with some protein. Electron micrographs of RNA'ase-treated cells showed the particles much flattened and linked together in rows. Their significance remains enigmatic.

Zaman (1961) reported that the characteristic uroid, or irregular tail region, of *Entamoeba invadens* consisted of a convoluted cyto-plasmic extension surrounded by dense vesicular material assumed to be a mucoid secretion. Ejection of digestive residues and excretion products into this mass was suggested.

The remarkable nuclear structure of *Amoeba proteus* was dis-cussed in Chapter 2. *Pelomyxa* nuclei lack the cortical honeycomb layer but have, just below the nuclear envelope, a zone of varying thickness containing a loose meshwork of fibrils 6 to 7 mμ in diameter (Pappas, 1959). Both genera have intriguing helical fibrils within the nucleus. The smaller amebae generally have conventional-looking nuclei. The sparse fibrogranular nucleo-plasm encloses clumps of chromatin and dense, usually peripheral, nucleoli. Only in *Endamoeba blattae* has the honeycomb layer of the *Amoeba* nucleus been found, but here on the external surface of the envelope. Since the genus *Endamoeba* is noted for its thick nuclear membrane in light-microscope preparations, other species may prove to have the honeycomb surface also.

In summary, it is evident that the free-living ameba, large or small, has a full complement of the protoplasmic organelles that are recognized as standard equipment for cells even of the highest organisms. The absence of restrictive morphological differentia-tions in the ameba has long endeared it to the experimentalist; the conventional nature of its cytoplasmic furniture may now reassure him that his pet is not unreasonably exotic. Such peculiarities as various species possess represent variations of these conventional structures that are more accessible to study than is the case with many other cells because of the ameba's extraordinary docility under surgery and other violent manipulations.

The remainder of the great array of rhizopods remains almost completely unexplored by electron microscopists. Most of them construct tests or shells around their bodies, a habit that is carried to an extreme by the foraminifera with their elegant, multi-chambered shells. These structures generally are mineralized and offer serious but surely not insurmountable obstacles to thin sectioning. The development of techniques for studying them should be well worth the effort for the light they may cast on phenomena of secretion.

A brief glimpse of the intricacies of foraminiferan shell structure is provided in a short report by B. Jahn (1953) of an electron-microscope study of shell fragments from recent and fossil species. The shell is pierced by pores that connect on the inner side to cylindrical tubes marked by regular annulae. These annulae are the sites of delicate, lace-like sieve plates extending across the tube lumen. It was not possible to relate these tubes in the isolated shells to the structure of the living inhabitant.

Some foraminifers and one group of testate amebae secrete shells that are primarily proteinaceous. A species of the latter group, *Gromia oviformis,* has very recently been subjected to electron-microscopic study by Hedley and Bertaud (1961). *G. oviformis* is a large marine species that occupies a spherical shell, about 10 μ in thickness, with an oral opening that is rimmed by a mucopolysaccharide capsule. During feeding and locomotion, a pseudopodial reticulum similar to that of *Allogromia* (p. 75) extends from the oral opening. Hedley and Bertaud were not able to observe pseudopodial structure in their material and directed their attention to the structure of the shell and enclosed cytoplasm.

The substance composing the proteinaceous shell is dense and homogeneous or sponge-like. It is regularly penetrated by radial canals containing traces of amorphous material. The oral capsule is composed of tubular fibrils, about 12 mμ in diameter, piled in layers. Within each layer the fibrils run roughly parallel to each other but at approximately right angles to fibrils in the adjacent layers. This arrangement is reminiscent of the alignment of collagen microfibrils in some vertebrate connective tissues, but no chemical similarity is suggested.

Lining the inner surface of the shell are several delicate lamellae with a remarkable structure. Each is composed of minute cylinders, about 10 mμ in diameter and 20 mμ long, their axes perpendicular to the plane of the sheet. Each cylinder is surrounded, at a center-to-center distance of about 21 mμ, by six others in a consistently orderly hexagonal arrangement. Filaments or septa interconnect the cylinders. As many as ten separate lamellae may be present, closely layered beneath the shell. Internal to these is a low-density zone of homogeneously granular material, considered to be ectoplasm, and a vesicular endoplasm. Cytoplasmic particulates include, in addition to Golgi elements and microtubular mitochondria, two kinds of solid bodies that are assumed to be fecal concentrates and concretions of metabolic waste products.

The chemical composition and the significance of the highly organized lamellae are completely unknown, and their disposition in the oral region was not determined. Similar packing of fibrils or granular bodies in hexagonal arrays is encountered in certain fibrous structures in ciliates (pp. 195, 202) and in a collagenous membrane in the vertebrate eye (Jakus, 1956), but there is no evidence that the packed units have anything more in common than meets the eye. Hedley and Bertaud suggest as a possibility that the lamellae are a unique sort of cytoplasmic surface membrane, since no other continuous boundary was found. Certainly the question merits further study.

Subphylum Actinopoda

In the Subphylum Actinopoda are to be found some of the most remarkable achievements in cellular architecture ever attained, expressed in the incredible elegance of the secreted internal skeleton. The siliceous skeletons of many of the Radiolarea and Heliozoea are delicately sculptured in designs of pure fantasy, while the acantharian skeleton exhibits an exquisite geometric precision, based on radiating spicules occurring always in multiples of ten.

Only two attempts have been made to examine skeletal structures, both utilizing spicules isolated from disintegrating heliozoan cells. Spicules of *Heterophrys marina* are not silicified; they are needle-shaped and show some evidence of internal structure (Wohlfarth-Bottermann and Krüger, 1954). The siliceous spicules of two species of *Acanthocystis* (Petersen and Hansen, 1960a) included plate-like and tack-shaped units of sizes and shapes characteristic of the species.

Studies of actinopods utilizing sectioning techniques are limited thus far to two related heliozoa that lack skeletons. *Actinophrys sol,* examined by Wohlfarth-Bottermann (1959b; also Wohlfarth-Bottermann and Krüger, 1954) is fairly small, with a single nucleus. *Actinosphaerium nucleofilum* (Text-fig. 4) studied by Anderson and Beams (1960) is much larger and multinucleate. In both, the spherical body contains a central mass of granular endoplasm surrounded by hyaline, vacuolated ectoplasm. As seen in electron micrographs, the ectoplasm both on the axopodia and on the body is conspicuously vacuolar, consisting of a veritable foam. Unlike the irregular vesicles filling the cytoplasm of *Amoeba* and *Pelomyxa,* most of the vacuoles in the heliozoa appear to be spherical, suggesting some turgor in the living state; some authors have suggested that such vacuoles are hydrostatic devices aiding in flotation. Most of them appear empty, but a class of small vacuoles observed abundantly in the axopodia of *Actinosphaerium* contains masses of dense particles (Fig. 30, Pl. IX). Anderson and Beams believe that these correspond to refringent bodies observable with the light microscope and suggest a comparison with the nitrogenous crystals of the giant amebae. In addition to the vacuoles, the ectoplasm of both genera contains mitochondria and some fine tubules and minute vesicles.

Axopodia differ from the slender, filose pseudopodia of some rhizopods in possessing an axial rod or thread that is birefringent in polarized light, surrounded by actively moving cytoplasm. Axopodia never form a reticulum; branching may occur but is inconspicuous. Contraction of the whole axopod is reported for some forms. Like the reticular pseudopodia of *Allogromia,* the axopodial surface is adhesive, and entraps particles that are carried along streaming paths. Jahn and Rinaldi (1959) suggested that shearing forces comparable to those postulated for *Allogromia* may operate here between the moving peripheral cytoplasm and the stationary axis.

The axial rods of the axopodia in both *Actinosphaerium* and *Actinophrys* consist of oriented bundles of packed fibrils embedded in a low-density matrix surrounded by ectoplasm. In the micrographs of *Actinosphaerium* (Fig. 30, Pl. IX) these appear as very fine filaments, 6 to 12 mμ in diameter and of indeterminate length. In *Actinophrys* they look tubular; they are roughly circular in cross-section, somewhat thicker than the filaments of *Actino-sphaerium,* and rather resemble the finer tubular elements of the cytoplasmic membranous reticulum. The authors of these two studies used different fixatives, at different pH's, and it is impossible to say at this point whether the difference between filaments and tubules is a real one or represents the effects of the fixing agents. Preservation of other components appears to be somewhat better in the micrographs of Anderson and Beams.

Neither of the pairs of authors offers any comment on the surface membranes of their cells. In Wohlfarth-Bottermann and Krüger's picture of an axopodium, the surface structure is discontinuous; this could be a fixation artifact. In the micrograph by Anderson and Beams, there appears to be continuous membrane following the contours of the surface vacuoles and covering the cytoplasm between. At the magnification provided, this seems to resemble the unit membrane of any other cell. This is of interest in view of Jahn and Rinaldi's speculations concerning the protoplasmic surface of the *Allogromia* pseudopodial reticulum. Since particles adhering to the surface move with the underlying cytoplasm, a continuous membrane over the two antagonistic streaming units is hard to imagine, yet no suggestion of a line of demarcation between them is detectable with the light microscope. None of the published illustrations of the heliozoa includes a complete transection of an axopodium, which may be taken to mean that these competent and experienced workers observed nothing startling in such micrographs.

In the cell body of the heliozoa, the axial filaments of the axopodia continue deep into the endoplasm. In *Actinophrys,* they insert directly on the membrane of the single nucleus. In *Actino-sphaerium* they are found close to the nuclei but apparently not directly connected to them. No centrioles or kinetosome-like structures were seen. The nuclei have double membranes, a finely granular nucleoplasm, and numerous dense, peripheral, small nucleoli. The endoplasmic matrix is fairly compact, containing abundant endoplasmic reticulum some of which bears Palade particles. Similar granules occur free of membranes. Also included in the endoplasm are microtubular mitochondria, Golgi bodies, food vacuoles surrounded by a multitude of small vesicles

PLATE IX

FIG. 30. Section through part of axopodium of *Actinosphaerium nucleofilum,* showing filaments of axial rod, canalicular endoplasmic reticulum, a mitochondrion at lower right, and vacuoles containing dense granules at top left. × 40,000. From Anderson and Beams, 1960.

FIG. 31. Section through young feeding stage of *Gregarina rigida,* partially embedded in host cell. Pellicular folds beginning to appear at free end, but no differentiation into ectoplasm and endoplasm, or into anterior and posterior body zones. Honeycomb layer of nuclear cortex lacking. Parasite cytoplasm has shrunk away from closely apposed membranes of host and parasite. × 20,000. From H. W. Beams and E. Anderson, unpublished.

FIG. 32. Section through contact surfaces of two individuals of *Gregarina rigida.* Note membrane layers, fibrils at tips of pellicle ridges, ectoplasmic network of fine filaments. × 50,000. From Beams, Tahmisian, Devine and Anderson, 1959a.

PLATE IX

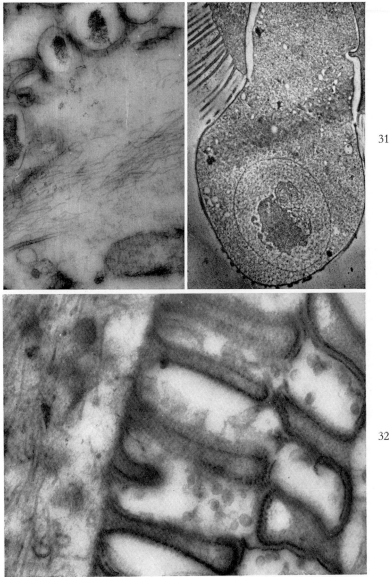

30

31

32

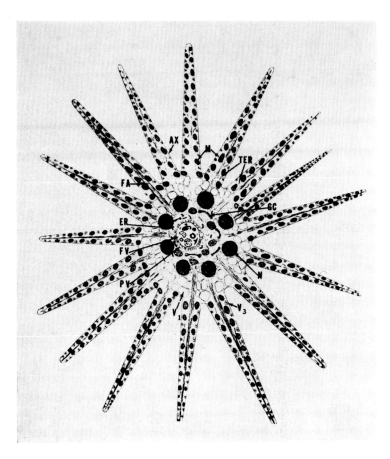

TEXT-FIGURE 4. Schematic drawing of *Actinosphaerium nucleofilum*.
AX, axopodium; ER, granular endoplasmic reticulum; FA,
filaments of axial rod; FV, food vacuole, GC, Golgi complex; M,
mitochondria; N, nucleus; PV, pinocytosis vesicles surrounding
food vacuole; TER, canaliculi of smooth endoplasmic reticulum;
V_1, non-contractile vacuoles of ectoplasm (not shown on axopodia);
V_3, vesicles containing dense granules. From Anderson and Beams,
1960.

suggesting pinocytic activity, and scattered empty vacuoles of moderate size.

Slime Molds

Most textbooks of protozoology state that the slime molds probably are not protozoa and then proceed to discuss them anyway. Not to break with tradition, we shall do likewise. Although in the formation of their multicellular fruiting bodies they show marked affinities with the fungi, earlier stages in their life cycle include flagellate or ameboid phases or both. In the cellular slime molds such as *Dictyostelium* and *Polysphondylium*, small amebae feed and divide repeatedly and eventually aggregate to form a multicellular mass or pseudoplasmodium, which typically assumes a sausage shape and subsequently migrates as a whole, although individuals can change their position within the mass. In the true slime molds or myxomycetes, such as *Physarum* or *Didymium*, flagellated stages may occur, followed by an ameboid form that by fusion and by nuclear division and growth without cytokinesis gives rise to a large, flat, spreading plasmodium in which active protoplasmic streaming occurs through a reticulum of veins or channels. A rhythmically alternating back-and-forth flow in the channels is often seen; otherwise movement in the solitary amebae and in the plasmodium is typically ameboid. Pinocytosis has been observed in *Physarum* (Guttes and Guttes, 1960). The phenomenon of protoplasmic streaming in the myxomycete plasmodium has attracted attention from students of ameboid movement, and biochemical studies have led to the extraction of a myosin-like protein, myxomyosin, by T'so, Bonner, Eggman, and Vinograd (1956).

Free vegetative amebae, aggregating stages, and pseudoplasmodia of *Dictyostelium discoideum* are the subject of a study by Gezelius (1961); the same species and *Polysphondylium violaceum* have been studied by Mercer and Shaffer (1960). No differences in cell structure are observed between solitary and aggregated cells. The cytoplasm resembles that of the small rhizopod amebae, with a rather dense fibrogranular matrix and vesicular elements of varying shapes and sizes. Some flattened membranous sacs have granular outer surfaces and in some individuals of *Dictyostelium* these granular membranes assume the form of extensive lamellae

;

(Mercer and Shaffer, 1960). Small spherical vacuoles are identified by the authors as contractile vacuoles, but these lack a structurally specialized cortex and are not associated with mitochondria. Mitochondria with typical twisted microtubules and a rather dense matrix are present, and occasional clusters of tiny vesicles suggest a resemblance to a Golgi apparatus. The nucleus is rounded, with a conventional double membrane and several dense nucleoli occupying dome-shaped bulges of the nuclear surface. Vacuoles, often with ruptured walls, that enclose bacteria in various stages of digestion are abundant in feeding and migrating amebae but disappear following aggregation. In these digestive vacuoles the bacterial cells become swollen and the cytoplasm disintegrates, but the cell walls remain and are surrounded by whorls of concentric dense membranes which ultimately appear to be ejected from the cell. Small, dense, homogeneous crystalline bodies appear in cells following aggregation.

The cytoplasm of the solitary cells is bounded by a distinct unit membrane. In the aggregate, cells are packed tightly together, with their slightly irregular surfaces separated by a rather constant, narrow (about 20 mμ) gap. Occasionally, small areas of higher density underlying the apposed membranes are seen, and Mercer and Shaffer pointed out a similarity with developing desmosomes of metazoan epithelial cells. However, no additional material is observed between the membranes, and permanent desmosomes would hardly be expected so long as the cells are free to shift positions within the aggregate.

Gezelius and Rånby (1957) and Mühlethaler (1956) investigated the structure of the fruiting body of *Dictyostelium* and found a cellulose stalk membrane with fiber structure similar to the cellulose of higher plants but of a lower crystalline order.

Electron micrographs of the plasmodia of *Physarum polycephalum* by Harris and Mazia (1959) and Stewart and Stewart (1959, 1961) agree in showing a rather dense fibrogranular cytoplasm without evident reticular membranes. Many clear vesicles of all sizes and shapes appear. Stewart and Stewart (1961) suggest that these vesicles may contribute membranes to the cell surface; an isolated droplet of extruded cytoplasm is seen in electron micrographs to have a highly convoluted surface membrane, underlain by a zone of homogeneous cytoplasm nearly empty of vesicles. Furthermore areas of presumptive cell membrane in plasmodia undergoing sclerotization (compartmentalization induced by drying or other rigors) are seen as rows of vesicles. Nuclei in *Physarum* are rounded and contain many discrete dark granules plus one to several round nucleoli. Mitochondria are microtubular.

Wohlfarth-Bottermann (1959a), studying plasmodia of *Didymium nigripes,* found that the presumably gelated areas are more compact in cytoplasmic texture and relatively poor in vesicular elements; internally such strands contain abundant fine tubules and vesicles, mitochondria, and a more diffuse matrix. In ripe spores the cytoplasm is extremely dense and uniformly granular; mitochondria appear unusually large and only a few vesicular structures with dense contents are present. In the germinating spore, the cytoplasm is loose in texture with many vesicles and abundant granular membranes. Wohlfarth-Bottermann's interpretation of these changes relative to sol-gel transformations is discussed above (p. 77).

Subphylum Sporozoa

The enormous assemblage of obligate symbiotes commonly referred to as sporozoa includes several probably unrelated series. Grassé (1952, 1953) places them in two subphyla: the Sporozoa, including the classes Gregarinomorphea, Coccidiomorphea, and Sarcosporidea, characterized by a vermiform, mobile sporozoite as the initial stage in the life cycle; and the Cnidosporidia, containing only three orders, whose initial developmental stage is an ameboid sporoplasm. A number of genera of undefinable affinities is loosely appended to the Sporozoa.

The existence of flagellated male gametes among the sporozoa is considered evidence of their flagellate ancestry; kinship with euglenoids, dinoflagellates, and bodonids has been suggested (see Grassé, 1953), but clear-cut evidence is lacking, as is to be expected after a long history of symbiosis.

In addition to their importance as agents of disease in man and domestic animals, the sporozoa are of considerable biological interest because of their remarkable adaptations to symbiosis: the complexity of their life cycles is matched by few other kinds of single-celled organisms and commonly includes stages in which they live exclusively within the cells of their hosts. Questions upon which studies of ultrastructure may be expected to throw significant light include the relationships of such intracellular forms to the host cells, and the morphological basis of a peculiar gliding locomotion without any visible means of propulsion, characteristic especially of gregarines.

Within the Subphylum Sporozoa, the Class Gregarinomorphea consists of relatively large symbiotes in the digestive tracts of invertebrates; often they have an intracellular phase early in the life cycle but typically they are extracellular at maturity. A thick, longitudinally striated pellicle is characteristic, and considerable internal differentiation is evident by light microscopy. The organisms examined to date with the electron microscope are all members of the Order Eugregarinida, and only vegetative individuals have been described. The most complete morphological studies are those by Beams, Tahmisian, Devine, and Anderson (1957, 1959a) on *Gregarina rigida,* by Kümmel (1958) on species of *Gregarina* and *Beloides,* and by Grassé and Théodoridès (1957, 1958, 1959) on species of *Gregarina, Stylocephalus,* and *Lophocephalus.* Shorter accounts are by Théodoridès (1959), Klug (1959), and Lacy and Miles (1959).

In most of the gregarines studied the body of the adult vegetative cell is covered with longitudinal crests or ribs about 1 μ high, formed by orderly folding of the cell surface (only in *Apolocystis,* examined by Lacy and Miles, the surface bears long, stiff hairs). Two limiting membranes are present, the inner one typically appearing thicker, the outer one often wrinkled. Both of these cover the pellicular folds. At least on the outer extremities of the crests and perhaps on their sides as well, fine fibrils run longitudinally or obliquely within the inner membrane (Fig. 32, Pl. IX). A third layer, thicker, less well-defined and variously reported as membranous or filamentous, forms a smooth, unfolded sheet running along the base of the crests in most micrographs but apparently is sometimes lacking. In the cytoplasm within the crests, Grassé and Théodoridès (1959) observed in some species compact groups of very fine fibrils.

Adhesion of potential gametocytes may occur in these gregarines, sometimes quite early in the vegetative phase. Pairs of organisms pictured in this state by Beams and colleagues show tip-to-tip contact of pellicular crests (Fig. 32, Pl. IX), but Kümmel and Klug both reported intimate interdigitation of crests.

Ectoplasm and endoplasm are more or less clearly distinguished by the absence from the former of inclusions such as granules of stored polysaccharide. Within the low-density ectoplasm beneath the ribbed cortex, a network of fine filaments may appear. Smaller ones, about 5 mμ in diameter, join to produce larger ones that connect with the surface envelope between crests. In addition the Beams group, found in micrographs taken at high magnification, larger, rather straight tubules that seem to terminate in pores at the cell surface. They suggest that these may be sites of secretion of mucus. Other authors have reported no evidence of pores. Circular myonemes are described by the Beams group and by Kümmel as bundles of very fine filaments lying a short distance beneath the cell surface. Their length and relationship with any other organelles are uncertain; whether they could be confused with the possibly filamentous layer underlying the pellicular crests is a moot point.

A typical feature of these gregarines is a distinct differentiation of the body into two parts, separated by a septum. The nucleus lies in the posterior part. The septum appears to consist of a substantial layer of fine filaments; Grassé and Théodoridès suggest that it is continuous with the layer beneath the pellicular crests, but no micrographs showing its peripheral connections have been published.

Golgi elements are reported by most investigators, but not by the Beams group. Microtubular mitochondria are shown by Kümmel in *Beloides,* and mitochondria of both microtubular and cristal types appear in the micrographs of *Stylocephalus* (Grassé and Théodoridès, 1959), but other workers have not found conventional mitochondria. It is possible that species differences are involved; the presence and absence of mitochondria in related species of the malaria parasite have been demonstrated (see below). Discrete arrays of parallel ergastoplasmic membranes are present near the nucleus in *Stylocephalus* (Grassé and Théodoridès, 1958, 1959). No evidence of phagocytosis in any of the feeding gregarines has been reported. The intriguing structure of the nuclear envelope has been described in Chapter 2.

Early developmental stages of vegetative *Gregarina rigida* are under study by Beams and Anderson (unpublished). These organisms undergo an attached phase during which they are partly embedded in the cells of the host gut. Electron micrographs (Fig. 31, Pl. IX) indicate a very intimate association between host and parasite cell membranes. The younger parasites lack pellicular ridges; these begin to develop on the free posterior surface of the body, where it protrudes from the host cell. Internal differentiations are minimal; the septum dividing anterior and posterior body regions is not visible, ectoplasm and endoplasm are not distinct, and no myonemes or fibrillar structures can be detected.

Most of the authors studying gregarine fine structure have been concerned with the mechanism of the eerie creeping movement of the adult cells. Body movements — flexion and twisting — are common, but gliding occurs in the absence of any detectable change in body shape. It is accompanied by the production of a conspicuous mucus trail, and this has prompted the suggestion that a sort of mucus jet propulsion operates. The consensus among recent students (Kümmel; Beams *et al*; Jarosch, 1959) is that mucus expulsion alone is not adequate to explain movement, but rather that a submicroscopic contraction of cortical structures causes local displacements of the body surface, perhaps against the mucus substrate. Kümmel, studying living cells, noted the simultaneous passage of particles over the body surface at different rates, independent of particle size, or in different directions. Given the semi-rigid longitudinal crests, it is conceivable that waves of contraction in cortical fibril systems or of minimal shortening of the circular myonemes, even though not visible at the light-microscope level, could result in body displacement.

For obvious reasons, the malaria parasite, *Plasmodium,* has been far more extensively investigated than any other sporozoan. Asexual generations occur in reptiles, birds, and mammals, sexual reproduction in mosquitoes. Sporozoites, inoculated into the vertebrate host by the bite of the insect, first enter phagocytic cells of the reticulo-endothelial system, and undergo one or more exoerythrocytic reproductive cycles, each mature cell producing many daughters called merozoites. The latter invade either more cells of the same type or erythrocytes. In the latter case repeated erythrocytic cycles follow, and eventually some of these merozoite develop into gametocytes which are infective for a mosquito. In the gut of the insect, asexual cells are digested but gametocytes produce gametes and fertilization occurs. The motile zygote invades the host tissue, grows, becomes multinucleate and produces many sporozoites which migrate through the mosquito some of them reaching the anterior digestive tract whence they may be injected into the vertebrate.

Electron-microscope studies of *Plasmodium* species parasitic in birds and mammals have been reported. Parasites during feeding and reproduction in the vertebrate are described by Fulton and Flewett (1956), Rudzinska and Trager (1957, 1959, 1961), Rudzinska, Bray and Trager (1960), Duncan, Street, Julian, and Micks (1959), and Meyer and Oliveira Musacchio (1960). Stages in the mosquito host have been examined by Duncan, Eades, Julian, and Micks (1960), Garnham, Bird, and Baker (1960), and Garnham, Bird, Baker, and Bray (1961). The greater part of the following account of vertebrate stages is drawn from the extensive, high-resolution studies of Rudzinska and Trager.

Several parasites may occur within a single host erythrocyte. They are embedded in the red cell cytoplasm and surrounded by two unit membranes, both of which are assumed to belong to the parasite (Figs. 33 and 34, Pl. X). (Duncan and colleagues, 1959) reported that some individuals were surrounded by three membranes and suggested that these were gametocytes.) The cytoplasm contains scattered or clustered Palade particles, variable amounts of endoplasmic reticulum and some larger vacuoles. Smooth-membraned sacs and vesicles identifiable as Golgi elements are seen occasionally. Mitochondria are present in several species of *Plasmodium,* but in *P. berghei* in rat blood (Rudzinska and Trager, 1959) none are evident. In this species elongate bodies composed of two to six concentrically arranged double membranes (Fig. 34, Pl. X) occur probably in all cells, and Rudzinska and Trager offer the suggestion that these may be the sites of oxidative activity. The membranes in some instances appeared to arise by invagination of both limiting layers of the cell surface. A second structure of unknown significance, seen in the same species, is a sausage-shaped vacuole with a double limiting membrane and structureless, low-density contents.

The most notable result of the studies by Rudzinska and Trager

PLATE X

FIG. 33. Section through two individuals of *Plasmodium berghei* in single host erythrocyte. Each contains large food vacuole filled with material resembling host cell cytoplasm. Food vacuole in cell at right indents nucleus. Assorted cytoplasmic membranes visible in both individuals. Small opaque granules enclosed in vesicles are residual hematin pigment. × 40,000. From Rudzinska and Trager, 1959.

FIG. 34. Section through part of *Plasmodium berghei* embedded in host erythrocyte. Note double membrane between host cytoplasm, below, and parasite. Lamellar body shown at center; tip of elongate food vacuole at left. × 60,000. From Rudzinska and Trager, 1959.

FIG. 35. Two dried whole cells of *Chromulina pleiades*. × 2,800. From D. R. Pitelka and E. C. Dougherty, unpublished.

FIG. 36. Section of *Chromulina psammobia*, showing kineto-rhizoplastic complex. See Text-fig. 6, showing the same structures in opposite (left-right) orientation. × 38,000. From Rouiller and Fauré-Fremiet, 1958a.

PLATE X

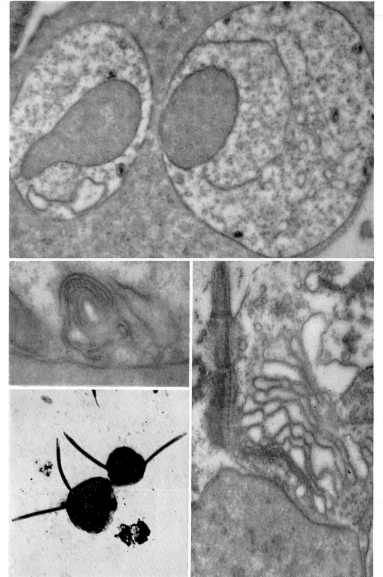

33

34

35

36

has been the discovery that cytoplasm of the host erythrocyte is taken into the *Plasmodium* body in food vacuoles invaginated and pinched off from the cell surface (Fig. 33, Pl. X). Except for minute bulges or wrinkles of the surface membranes, the contours of the cell are smooth and even, with no suggestion of ameboid activity; hence this drawing-in of food is a rather novel form of phagocytosis and suggests a membrane activity characteristic of pinocytosis. Digestion of the ingested protoplasm, with concomitant accumulation of the residual pigment, hematin, occurs either within the food vacuole as in *P. lophurae* in duck red cells, or in subsidiary vacuoles budded off from it in *P. berghei*. Food vacuoles frequently occur near the rather large nucleus and may indent it deeply. The growing parasite eventually engulfs most of the erythrocyte cytoplasm, and nuclear fissions begin. At this stage, studied in *P. lophurae,* the endoplasmic reticulum is abundant and mitochondria are numerous. Double membranes begin to appear in the cytoplasm, outlining areas into which the cytoplasmic organelles are segregated and which develop into individual merozoites. The remainder of the mother cell, consisting mainly of old food vacuoles and lipid droplets in a watery fluid, is left behind as a residual body when the merozoites are completed. No mitotic events were observed.

Studying exoerythrocytic stages in the chicken parasite, *P. gallinaceum,* in infected tissue cultures, Meyer and Oliveira Musacchio reported finding no mitochondria and no evidence of food vacuole formation in feeding stages. The cytoplasm contained many dense bodies that were not identifiable at the resolutions achieved. Merozoite formation proceeded as described by Rudzinska and Trager. Liberated merozoites often had a dense, ring-like structure just beneath or in the cell membrane at one pole.

This ring- or cup-shaped structure, about 120 mμ in diameter, was also found at one pole of sporozoites of *P. gallinaceum* and of *P. falciparum* by Garnham and colleagues (1960, 1961), who studied them in the salivary glands of infected mosquitoes. In addition these authors found a paired organelle consisting of two dense, flattened, elongate (about 1·4 μ) lobes narrowing at the anterior end in proximity to the polar cup, and 12 or more tubular fibrils about 19 mμ thick, arranged longitudinally at the periphery of the cell under a rather thick pellicle consisting of two membranes plus a granular intervening layer. A lateral depression in the pellicle, bordered by a thickened rim, is present in the *P. falciparum* sporozoite and is termed a micropyle. Garnham and colleagues suggest that after penetration of a cell in the vertebrate host, the sporozoite discards its differentiated pellicle by emerging through the micropyle.

Duncan and colleagues examined sections of mosquito stomachs at increasing intervals after their infection with *P. cathemarium* from canaries. Beginning with the third day, the progressive development of sporozoites within the encysted zygotes (oocysts) could be observed. The oocyst is embedded between epithelial cells of the host's stomach and is surrounded by a thick, homogeneous capsule that is limited internally by the parasite's cell membrane and externally by the membranes of neighboring host epithelial cells. The material of the capsule is indistinguishable from the elastic layer of the stomach wall and may be continuous with it. As development progresses, the capsule becomes thinner and forms abundant protrusions into the oocyst; ultimately it is no longer visible as a continuous layer.

The early oocyst cytoplasm is extremely dense, so that internal organelles cannot be distinguished. At five days it is somewhat looser in texture, vacuoles appear, and many nuclei, nucleoli, and mitochondria can be distinguished. The size of the oocyst increases greatly during development; the authors consider that the decreasing compactness of the cytoplasm in later stages cannot account for this and that materials for protein synthesis as well as water must enter the parasite via the capsule. By seven or eight days after infection, the oocyst cytoplasm is segmenting into thousands of sporozoites. Mitochondria are long and branched, endoplasmic reticulum is present, and Palade particles are found on the reticulum or scattered in the matrix. By ten days sporozoite development may be complete. Released sporozoites often are embedded in host cells, but separated from their cytoplasm by host cell membrane as well as their own double surface envelope. Their cytoplasm is moderately dense and contains, in addition to the usual organelles, rod-like bodies, typically in pairs or perhaps triplets, which presumably correspond to the paired organelle described by the Garnham group.

These accumulated observations cover most stages in *Plasmodium* development except for the sexual phases in the mosquito and the young motile zygote. It appears that from early developmental stages the parasite is enveloped by at least two unit membranes. In the vertebrate erythrocyte it clearly is embedded in host cytoplasm, probably with no intervening host membrane. Feeding by ingestion of host cytoplasm in the erythrocytic stage is established; in extraerythrocytic parasites further study is needed. The formation of limiting membranes around developing merozoites and sporozoites may perhaps proceed by invagination of sheet-like extensions from the parent cell membrane, or it may occur by neoformation of membranes within the cytoplasm. This question has not been resolved, but in any event the increase in membrane area is enormous. The fact that some species of *Plasmodium* possess morphologically typical mitochondria while others lack them poses a problem of considerable interest. Trager (1960) recently has been able to maintain the bird parasite *P. lophurae* apart from its host erythrocytes for a limited time, permitting a significant study of nutritional requirements in this form and offering a promising approach to the study of such matters as host specificity. Metabolic studies on species with and without identifiable mitochondria might, hopefully, be possible by these means.

Sporozoites and merozoites are supposed to be motile, at least briefly. Invasion of reticulo-endothelial cells apparently is accomplished in part by the host's phagocytic activity (Meyer and Oliveira Musacchio, 1960), but entrance into erythrocytes — which has only rarely been observed in the light microscope (Trager, 1960) — must be an active process. Enzymes are probably released to facilitate penetration but it seems as though some mechanical activity must be involved. Garnham and colleagues were tempted to suggest that the peripheral fibers seen by them in sporozoites could be related to locomotion, while the paired organelle was a glandule secreting lytic enzymes and the anterior ring might function in attachment. Only the polar ring has been described in any stage of the vertebrate cycle. All three of these enigmatic features are suggestively similar to structures seen by several authors in *Toxoplasma* and in *Sarcocystis* (see below).

Babesia is an organism parasitic in vertebrate erythrocytes and with a life cycle also involving an arthropod host. It is recognized by Grassé as a sporozoan but not allocated to any recognized class. It may be noted here that in an unillustrated abstract Rudzinska and Trager (1960) describe it as being quite similar to *Plasmodium* in ultrastructure. Like the latter it apparently ingests host cytoplasm, but it does not produce hematin-like pigments as a digestive product.

The Class Sarcosporidea contains the single genus *Sarcocystis,* of uncertain affinities and incompletely known life cycle. It is a common parasite in the striated muscles of mammals, birds, and reptiles, where large cysts are found, containing tremendous numbers of banana-shaped organisms. Ludvik has used electron microscopy to study *Sarcocystis tenella,* parasitic in the sheep (1956, 1958), and *S. miescheriana* from the pig (1960). Except for the structure of the cyst wall, the two species are very similar, and Ludvik's interpretation is summarized in Text-fig. 5. The body is covered by a double membrane, which is thickened at the anterior apex to form a polar ring. Within or beneath the pellicle 22 to 26 fine fibrils radiate from the polar ring and pass to the posterior end of the cell. Immediately beneath the polar ring is an internal organelle called the conoid, a truncate, funnel-shaped, dense structure. The anterior third of the body is filled with parallel longitudinal fibers, moderately dense and about 50 mμ in diameter, arranged in orderly rows. Behind this zone is an accumulation of large, dense, spherical granules of unknown composition. The posterior third of the cell contains a granular cytoplasm enclosing the nucleus, polysaccharide granules in vesicles, clear vacuoles, and several large mitochondria. Since growth as well as reproduction occurs within the cyst, exchange of material with the host muscle tissue must be possible. In the sheep parasite the outer layer of the cyst wall is spongy, the inner one broader and uniformly finely granular. The wall of the pig cyst consists of a rather thin layer of fibrogranular material resembling cytoplasm, and long villi containing fine, wavy, longitudinal tubular fibrils in a low-density matrix. These villi penetrate the host muscle tissue to a depth of some microns.

Among the parasitic organisms supposed to be protozoa but unclassifiable on present evidence, none has received more attention than *Toxoplasma* (see Ball, 1960). Unlike any other

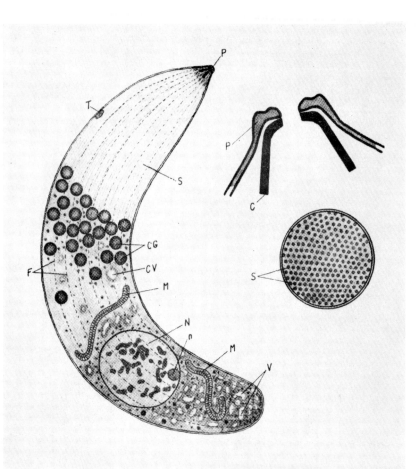

TEXT-FIGURE 5. Schematic drawing of *Sarcocystis tenella*. Polar ring and conoid enlarged at upper right, cross-section through anterior body region at lower right. C, conoid; CV, unidentified low-density vesicles; CG, dense granules of the central body zone; F, longitudinal fibers beneath the pellicle; M, mitochondria; N, nucleus; n, nucleolus; P, polar ring; S, fibers filling the anterior third of the body; V, vesicles in the posterior cytoplasm. From Ludvik, 1958.

protozoan parasite, its lack of host specificity is so extreme that many workers recognize only a single species occurring in a wide variety of birds and mammals and in various tissues within the host. Electron microscopy (Gustafson, Agar, and Cromer, 1954; Meyer and Andrade Mendonca, 1955; Ludvik, 1956) reveals a set of characteristic anterior organelles: a polar ring and conoid like those of *Sarcocystis,* and extending posteriad from within the conoid a group of 14 to 18 longitudinally oriented, claviform threads, 80 to 120 mμ in diameter, called toxonemes. In addition, longitudinal pellicular fibers diverge from the polar ring. The cytoplasm more posteriorly is of conventional appearance and contains bodies believed to be mitochondria, in addition to a nucleus with a distinct membranous envelope. Lainson, Baker, Bird, Garnham, and Healey (1961) report that *Lankesterella* (placed with the coccidiomorphs in the Grassé treatise) resembles *Toxoplasma.*

The similarity in structure of the anterior organelles in *Sarcocystis* and *Toxoplasma* is too great to be ignored, and it seems likely that taxonomic rearrangements are in order. It is unquestionably premature to draw similar conclusions with respect to these two organisms and *Plasmodium*; but the apical ring, peripheral fibers, and paired organelles variously reported by the Garnham group, the Duncan group, and Meyer and Oliveira Musacchio invite further comparative study.

Anaplasma is yet another infectious organism, whose classification as a protozoan has seemed highly dubious. Recent electron-microscope studies (Foote, Geer, and Stich, 1958; Espana, Espana, and Gonzales, 1959) leave the question still open. Foote and colleagues sectioned infected bovine erythrocytes and found wagon-wheel-shaped structures that seemed to be particulate in nature; they concluded that the organism is probably a virus. Espana and co-workers, however, found in particulate preparations of hemolyzed erythrocytes an opaque, spherical head, up to 1 μ in diameter, and a long, curved tail; these often were associated in pairs as rings, dumbbells, or open triangles. According to these authors, motility can be detected by phase-contrast study of living organisms.

Members of the Subphylum Cnidosporidia develop their ameboid sporoplasms within a spore that also contains one or more coiled polar filaments, typically enclosed in capsules, that may be extruded under certain conditions. The spore wall often is multicellular and this feature, among others, has led some authorities to propose that the cnidosporidians are degenerate metazoa. Electron-microscope studies include a brief observation of whole spores of *Plistophora* by Steinhaus (1951) and two papers concerned with species of the genus *Nosema* (Weiser, 1959; Huger, 1960), parasitic in the fat bodies of insects. Both genera belong to the Order Microsporidia, characterized by less complex spores than many other cnidosporidians.

Huger's micrographs show that the *Nosema* spore wall has two components: a thick, homogeneous inner layer and a lamellar outer layer possibly derived from host tissues. The inner coat thins at the site of the prospective pore. Within the wall is a low-density anterior body with an apparent laminated structure. It is not a true polar capsule; Huger calls it a polarblast and suggests that it is a substance capable of swelling to cause extrusion of the polar filament. The latter inserts in or near the prospective pore, runs straight through the polarblast, then turns to the periphery of the spore and spirals through as many as 18 turns. Its diameter tapers from 230 mμ at its insertion to about 70 mμ at its free posterior tip. It is enclosed in a double membrane and contains longitudinal fibrils in a rather dense matrix. Because of this density, observation of detail in the fibril bundle is difficult but Huger notes a suggestive similarity to the organization of a flagellum. The situation needs to be reexamined since there is strong light-microscope evidence that the polar filament is a tubule that everts upon extrusion (Kramer, 1960).

Clusters of dense granules, probably polysaccharide, occur at one side of the polarblast. The cytoplasm in the posterior part of the spore is granular and too dense to reveal much internal structure; two even denser nuclei are seen.

CHAPTER 4

PHYTOFLAGELLATES

IN THIS section, "Phyto" will be de-emphasized in favor of "flagellates". Any consideration in depth of such plant-like features as chloroplasts and cell walls is beyond the scope of this review and the competence of its author, and those aspects of phytoflagellate structure that are particularly significant to a study of the evolution of the higher algae and land plants must be largely neglected. Rather, we are interested in the phytoflagellates as spectacularly successful single-celled organisms whose ancestors presumably existed for long ages before the first entirely animal-like protozoan groups split off.

One would like to be able to turn with some certainty to a protistan group that includes the most primitive eucells in existence today, and to ask what is the minimum structural equipment with which such a cell can survive as an independent organism. Unfortunately, not only is it improbable that any contemporary organism exemplifies an ungarnished minimum cell organization, but we do not know which of the several distinct phytoprotistan lines is of most ancient origin, hence likely to include the most nearly primitive types.

If a monophyletic origin of all eucells is postulated, then particular attention must be paid to the non-flagellated red algae, which synthesize pigments more nearly like those of the blue-green algae than do any other eucellular protista. Thus Dougherty (1955) and Dougherty and Allen (1960) offer the unconventional proposal that the first flagellates evolved from already eukaryotic cells resembling simple red algae.

An electron-microscope study has been reported by Brody and Vatter (1959) of the unicellular red alga, *Porphyridium cruentum*. This small cell is almost entirely occupied by a single chloroplast, which contains the characteristic biliprotein pigment as well as chlorophyll. The shape of the chloroplast is irregular, and the

101

internal lamellae occur as single discs, not bands of discs, following parallel contorted paths through the plastid matrix. As noted in Chapter 2, this arrangement is more like that in the blue-green algae than those of any other algae (Gibbs, 1960). A central pyrenoid is traversed by a few sinuous lamellae, but is not associated with starch grains which appear instead in the cytoplasm outside the plastid, each grain surrounded by a membrane. In the peripheral cytoplasm are assorted small membranous structures identified as endoplasmic reticulum, and stacks of smooth lamellae resembling Golgi elements, but no mitochondria. The nucleus has a dense eccentric nucleolus and is surrounded by a double membrane. The cell is bounded by a plasma membrane and enveloped by a thick sheath with a finely fibrous texture.

The absence of mitochondria in an aerobic, photosynthetic, nucleated cell may be a distinctly primitive feature; bacteria and blue-green algae lack mitochondria, and there is evidence to suggest that mitochondrial enzymes in them are localized in the cell membrane or elaborations of it (Robinow, 1960). A study of the sites of oxidative activity in *Porphyridium* would seem to be in order. Otherwise, *Porphyridium* appears to have a minimum complement of all the conventional eucellular equipment. It may be morphologically the simplest free-living eucell yet investigated.

Interestingly enough, the red algae, which never bear flagella at any time, also appear to lack centrioles, although reinvestigation of a few reports of centriole-like granules in the literature of light microscopy (see Fritsch, 1945) is needed. Unless they have secondarily lost the kinetosome/centriole, we must assume either that they developed nuclear membranes, chloroplasts and Golgi structures independently of all other eucellular stocks, or else, with Dougherty and Allen, we may suggest that the modern reds descended from a *Porphyridium*-like archetypal eucell that also was ancestral to other higher protists.

Once the kinetosome/centriole evolved, diversification of flagellate groups occurred on a prodigious scale. Perhaps this was coincidental, or was consequent on the extent of dispersal and subsequent isolation permitted by flagellar motility. But the variety of intracellular structures that, as we shall see, are intimately associated with kinetosomes suggests that this new organelle provided an extraordinarily adaptable means of enhancing structural organization.

Among the phytoflagellates, diversity and relationship are most readily assessed biochemically; thus the kinds of photosynthetic pigments, of products of photosynthesis, and of skeletal or wall components are key taxonomic characters. Purely morphologic features such as number and position of flagella and nuclear cytology generally are less reliable though still often significant indicators. Except for some remarkable surface specializations, anatomy at the cellular level does not become as elaborate as it does in the zooflagellates. The number of phytoflagellates thoroughly investigated with the electron microscope is still so small that we cannot say whether ultrastructural differences can be correlated with biochemical peculiarities; nonetheless there are a few promising clues.

The phytoflagellates may be roughly divided into two large assemblages, the brown stock and the green stock, characterized by different sorts of pigments. Inviting lines of biochemical research are suggested by speculations on the phylogenetic relationships between and within these groups (see Hutner and Provasoli, 1951; Dougherty and Allen, 1959, 1960). We shall consider the brown stock first. Several important groups within these assemblages have yet to be studied with the electron microscope.

Class Chrysomonadea

The Chrysomonadea include many familiar and conspicuous genera, some of which have been used widely in axenic culture for biochemical studies. They also contribute importantly to the indescribably huge numbers of cells that constitute the microplankton of marine and fresh waters. Most of these are poorly known because of their minute size and fragility. They escape an ordinary plankton net, and only the most dedicated light microscopists have been willing to bother with them at all. In a notable series of recent studies, Parke, Manton and Clarke (1955–1959) and Manton and Leedale (1961a, 1961b) have applied combined techniques of light and electron microscopy to the abundant marine planktonic species of the genus *Chrysochromulina*. Thanks to this and other work, it is becoming apparent that the

H

morphological criteria applied to the classification of groups within the Chrysomonadea need to be redefined or in some cases discarded. Primarily these have been based on numbers and relative lengths of flagella. Electron microscopy indicates at least that flagellar ultrastructure must be considered as well.

One of the first results of electron-microscope study of protozoa was the confirmation of light-microscope observations of the existence of lateral filaments or mastigonemes on the flagella of certain kinds of protozoa. Meticulous studies by Fischer in 1894, by Mainx in 1928, by Petersen in 1929, by Deflandre in 1934, and Vlk in 1938 (see Brown, 1945, and Pitelka, 1949, for descriptions and citations) had shown that four basic types of flagella could be distinguished. These, following Deflandre's terminology, are: simple flagella, with no filamentous appendages; acroneme flagella, with a single terminal thread of variable length and diameter; pantoneme flagella, with mastigonemes along both sides of a flagellum seen in profile; and stichoneme flagella, with a single row of mastigonemes. Much argument over the reality of these filaments ensued, but the facts that pantoneme flagella were observed rather consistently in chrysophycean and xanthophycean flagellated cells and stichoneme flagella in euglenoids lent considerable weight to the view that they were not mere artifacts. Electron-microscope studies by Brown (1945), Pitelka (1949), Chen (1950b), Houwink (1951, 1952), Brown and Cox (1954), Pitelka and Schooley (1955), and especially Manton and her colleagues (1951 *et seq.*) now leave no room for doubt that they are real filaments with definite structure and that, at least within certain groups, their occurrence is so regular as to constitute a valid taxonomic character.

In the motile unicellular species and the flagellated zoospores and spermatozoids of multicellular species of some of the Chrysophyceae, Xanthophyceae, Phaeophyceae, and three groups of aquatic fungi, two unequal flagella per cell are common. One of these, usually longer and directed forward in locomotion, is pantoneme; the second, shorter and trailing, is simple or acroneme. A typical pantoneme flagellum is that of *Ochromonas malhamensis* shown in Fig. 37, Pl. XI. The site of attachment of mastigonemes to the flagellum has not been determined with certainty; they are too fine to be followed in most sectioned material and are likely to be distorted or detached in whole-mount preparations where flagella have undergone enough disintegration to reveal internal structure. However, the weight of evidence has led Manton (1955a, 1959b) to conclude that they arise from two specific fibrils of the peripheral nine in the flagellar axis.

The significance of mastigonemes is obscure. Their presence on the flagella of still living though moribund cells has been observed by dark-field light microscopy by Vlk (1938) and by Pitelka (1949). When they are relatively rigid, as their appearance in fixed cells such as Fig. 37, Pl. XI might suggest, they may serve to increase the surface area of the flagellum and hence its mechanical effectiveness. This suggestion gains some support from the fact that it is the more active flagellum of the heterodynamic pair that is pantoneme; furthermore, in the spermatozoid of the brown alga *Dictyota* (Manton, 1959b), where the second flagellum is repre- sented only by a basal body, the single functional flagellum is pantoneme.

On many flagella, filaments much finer than the mastigonemes may appear, mingled among them as in Fig. 37, Pl. XI, forming a fine feltwork along the flagellar surface (as in some euglenids, see Fig. 42, Pl. XII), or sparsely arrayed along simple or acroneme flagella. These filaments have been observed frequently, but nothing is known of their distribution or significance.

The pantoneme flagellum as it is seen in these brown-stock algal groups is a clearly defined type, characteristic of cells that have two unequal heterodynamic flagella. In other chrysomonads, however, the two flagella may be equal or nearly so in length and homodynamic. Still others are reported to bear three flagella, or only one. Most of these flagella are simple or acroneme, and electron-microscope studies prove that distinctions based strictly on numbers of flagella are shaky at best, as we shall see.

Before undertaking a review of chrysomonad internal anatomy as revealed in sectioned cells, we may consider briefly some additional electron-microscope observations on superficial struc- tures. In many chrysomonads, cellulose or pectin occurs in a thin, supple cell wall. In addition, calcareous, siliceous, or organic scales are common, irregularly disposed over the surface or aligned with great order; sometimes these bear remarkably long, articulated spines. Scales have been studied at length with the

PLATE XI

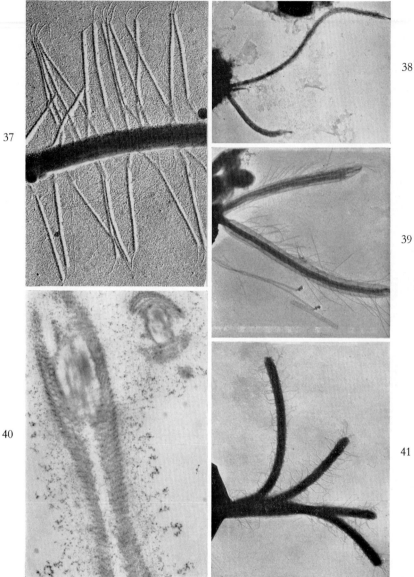

light microscope, but electron microscopy reveals their exquisitely varied detail and has permitted taxonomists to distinguish specific types, both by study of whole detached scales and by preparation of replicas reproducing the contours of intact cell surfaces. Thus for example the delicately scalloped scales of *Chrysosphaerella* have been studied by Fott and Ludvik (1956a) and Harris and Bradley (1958), the peculiar, bristled, thumb-tack scales of *Physomonas* (= *Paraphysomonas*) *vestita* by Houwink (1952), the ribbed and pitted scales of *Synura* by Manton (1955a) and the tiny, elegantly sculptured scales of *Chrysochromulina* by Parke, Manton, and Clarke (1955–1959). Generic revisions are based in part on extensive studies of scale variations in *Mallomonas* by Asmund (1955, 1959) and Harris and Bradley (1960) and in *Synura* by Petersen and Hansen (1956, 1958). Deflandre and Fert (1953) and Parke and Adams (1960) have examined the scales of coccolithophorid flagellates, a group closely allied to the chrysomonads.

Additional light on the question of scale structure is provided by the micrographs of sectioned *Chrysochromulina* (Parke, Manton, and Clarke, 1958, 1959; Manton and Leedale, 1961a) and *Paraphysomonas* (Manton and Leedale, 1961c). In each of the *Chrysochromulina* species illustrated in these papers, two kinds of scales occur, differing from each other in size, shape, and sculpturing. In *C. ericina* some scales are flat two-layered discs and others have enormously long spines. In *C. chiton,* larger scales are saucer-shaped and smaller ones alternate with them, lying beneath and overlapping the rims of two adjacent saucers. In *C. strobilus,* the larger scales form a close mosaic over the cell membrane and the smaller scales lie above these. Thus the problems of scale production and disposition are even more complex than the elaborate patterns of the individual scales might suggest. Production of scales in intracellular vesicles is demonstrated in *Paraphysomonas vestita* (and also in a green alga, p. 120).

Usually considered to be the most primitive chrysomonads are the supposedly uniflagellate chromulinids. The basis for this distinction became suspect with the publication of a detailed study of *Chromulina psammobia* by Rouiller and Fauré-Fremiet (1958a) showing the presence of a second flagellum (*Chromulina pleiades,* examined briefly by Pitelka and Dougherty [unpublished] some years ago, has two equal acroneme flagella [Fig. 35, Pl. X]. The species formerly called *C. pusilla* is a true uniflagellate but it has been transferred to the green algae and to the new genus *Micromonas* by Manton and Parke [1960].). The ultrastructure of *C. psammobia* may be described here as generally typical of chrysomonads (Text-fig. 6).

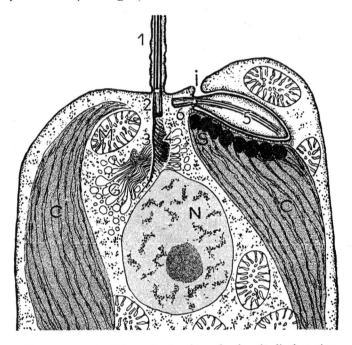

TEXT-FIGURE 6. Schematic drawing of a longitudinal section through the anterior half of *Chromulina psammobia*. C, chloroplasts; G, Golgi body; I, invagination of the cell surface enclosing the internal flagellum; N, nucleus; M, mitochondrion; S, stigma. The kineto-rhizoplastic complex includes 1, the free flagellum; 2, its kinetosome; 3, rhizoplast; 4, dense zone tentatively identified as a centrosome. The kineto-stigmatic complex includes 5, the internal flagellum lying above the stigma and 6, its kinetosome. From Rouiller and Fauré-Fremiet, 1958a.

The cell body, up to 15 μ in length, is limited by a typical unit membrane, without any clearly evident outer coating or wall. The cytoplasm is fibrogranular and encloses numerous mitochondria of typical protozoan microtubular structure, assorted small vesicles referable to the endoplasmic reticulum, and lipid and leucosin droplets. The two elongate lateral chloroplasts contain more or less flexuous lamellae, oriented longitudinally and converging toward the two extremities of the plastid. At the anterior end of one chloroplast is the stigma, a curved plate of ovoid vesicles packed in a single layer and containing a dense homogeneous substance; the walls of the vesicles are continuous with the chloroplast lamellae. Occasional cells show secondary vesicular zones at the posterior end of one or both chloroplasts. No pyrenoids are described, but one of the published micrographs shows a zone in one chloroplast in which the matrix is quite dense and the lamellae regularly spaced, resembling a pyrenoid of the *Euglena* type (p. 23).

The nucleus of *Chromulina psammobia,* located between the chloroplasts in the anterior half of the cell, has a conventional double membrane and central nucleolus. The single, long, motile flagellum inserts at the anterior end of the cell, where the central pair of flagellar fibers originates at a small, dense, flat granule at the level of the cell surface. From one side of the base of the conventional fibrous kinetosome a slender, obliquely striated ribbon extends to the cone-shaped apex of the nucleus (Fig. 36, Pl. X). No direct contact between this ribbon and the nuclear membrane is evident, rather the ribbon bends and may run a short distance along the nuclear surface before disappearing. The period of the striations is about 42 mμ. Fibers called rhizoplasts, passing from flagellar basal bodies to the nuclear surface, have been described in many flagellates by light microscopists. The ribbon-like filament of *C. psammobia* is a classical example of a rhizoplast seen at the electron-microscope level; however, it will be seen that very similar fibers in other flagellates may have no detectable association with the nucleus, at least in the non-dividing stages examined.

On the medial side of the rhizoplast is an elongate, dense mass that accompanies it almost to the nuclear surface. Rouiller and Fauré-Fremiet (1958a) suggest that this may be a centrosome, but no observations of mitotic stages were available. A well-defined Golgi zone appears in a constant position between the kinetosome and the nucleus, just lateral to the rhizoplast.

Close to the kinetosome of the external flagellum, on the side opposite the point of origin of the rhizoplast, is a second kineto-some, perpendicular to the primary one and parallel to the cell surface. From its apex, a very short (about 1 μ) flagellum extends into a tubular invagination of the cell surface. This second flagellum, lying in its pit and thus surrounded throughout its length by a flagellar membrane and by the invaginated cell membrane, occupies the concavity of the stigma. The flagellum contains the usual 11 fibrils in an unusually dense and expanded matrix, so that the whole organelle is spindle-shaped.

Rouiller and Fauré-Fremiet refer to the two kinetosomes and their associated structures as the kineto-rhizoplastic complex and the kineto-stigmatic complex. Both complexes occur in other flagellates. The kineto-stigmatic complex has been discussed in Chapter 2.

In a brief report illustrated only by a diagram of a non-flagellated cell, Hovasse and Joyon (1957) describe the ultrastructure of a colonial chromulinid, *Hydrurus foetidus*. Cells of this species secrete a tough gelatinous substance that binds the division products together until long, multibranched palmelloid colonies are formed. Recently divided cells may transform into supposedly uniflagellated swimmers and leave the colony. The non-flagellated colonial cell, according to Hovasse and Joyon, shows clear polarity. The nucleus, Golgi zone, and leucosin droplets are found in the anterior part of the cell, the large chloroplast posterior to these, with microtubular mitochondria scattered throughout. Many clear, small, vesicles under the cell surface probably are mucigenic bodies that secrete the palmella jelly. No kinetosome is indicated in the diagram and the authors do not comment on the process of flagellum production. They refer to a single flagellum in the swimming cell, and remark that the Golgi apparatus is more distinctly oriented in the flagellate than in the palmelloid state.

Although *Ochromonas* and related small species of the biflagellate family Ochromonadidae are cultured in chemically defined media and available in laboratories everywhere, no electron-microscope studies of the type genus have yet been published. Two other ochromonads, *Synura caroliniana* (Manton, 1955a) and *Para-physomonas vestita* (Manton and Leedale, 1961c) have been studied in detail with the electron microscope. The spherical colonies of

Synura are composed of small cells with elongate stalks radiating from a common center. Body and stalk are covered with overlapping siliceous scales shaped like oval hat brims with narrow peaked crowns. The scales clearly are superficial; they lie outside the cell membrane and contain no protoplasm. The stalks contain low-density protoplasm continuous with that of the cell bodies. In the ovoid cell, the two large, curved plastids are symmetrically placed with the nucleus and a large leucosin vacuole between them. No pyrenoids or stigma are observed. Microtubular mitochondria are scattered through the cell but more numerous near the basal bodies of the two flagella. The latter are of conventional structure, with a transverse septum occurring at the level of the cell surface where the central flagellar fibrils appear. From the base of at least one of the kinetosomes, two or more obliquely striated fibers, like the rhizoplast of *Chromulina,* pass posteriad to the conical apex of the nucleus, where they branch out and spread along the nuclear surface like an "enveloping string bag". Other slender, unstriated roots apparently pass from the kinetosomes upward and outward to unknown destinations.

Again as in *Chromulina,* a distinct Golgi apparatus is found in a constant location beside the rhizoplasts. Of the two flagella, the longer, pantoneme one has an expanded sheath within which the axial fiber bundle is eccentrically placed. Where the planes of symmetry of the two flagella are discernible near the body of the organism, they appear to be identical; that is, a single plane bisects both flagella.

Paraphysomonas vestita is a non-pigmented relative of *Ochromonas* and is efficiently phagocytic — a rather neat trick for a cell that is covered with long-spined scales. General cytoplasmic organization is similar to that of other chrysomonads. From the sides of the two anterior kinetosomes, several bundles of cross-striated fibrils pass down to the adjacent, pointed, anterior tip of the nucleus and along its sides. Nearby, on one side of the nucleus, lies the extraordinarily large, well-developed Golgi body, with extensive flat sacs and abundant peripheral microvesicles. The contractile vacuole is identifiable and has been described in Chapter 2.

Poteriochromonas stipitata, a solitary ochromonad that, like many of its relatives, indulges in both photosynthesis and phagotrophy, was briefly examined in thin section by Wolken and Palade in 1953. Although the high quality of the single micrograph published was notable for that early date, it does not reveal any morphological features peculiar to the genus. The chloroplasts have rather uniformly parallel lamellae and flank the anterior nucleus. Neither rhizoplast nor Golgi bodies were noted by the authors. In an unidentified solitary ochromonad, possibly *Ochromonas,* occurring as a contaminant in a bacterized ciliate culture studied by Pitelka (unpublished), Golgi elements were seen in a constant position anterior to the nucleus and obliquely posterior to the two flagellar basal bodies, but observations were too few to justify any conclusions regarding the presence or absence of a rhizoplast. The chloroplasts in this organism appeared exceptionally small, and the posterior cytoplasm was filled with vacuoles containing bacteria or their partially disintegrated remains.

A brief study of whole mounts of the species *Stokesiella dissimilis* (Petersen and Hansen, 1960b) should be mentioned here. The genus often is placed among the lower zooflagellates (see Bicoecida, p. 136), but Hollande (1952a) recognizes it as a member of the family Ochromonadidae. Like many chrysomonads, the organism dwells within an open sheath or envelope produced by the cell. In Petersen and Hansen's figures only the longer flagellum was visible emerging from the envelope, and this was clearly of the ochromonad pantoneme type. The delicate, cup-shaped envelope consisted of a single band of striated membranous material, wound in a tight spiral with thickenings at its over-lapping edges.

Chrysochromulina is one of a group of genera considered to be triflagellate. One of the most interesting results of Parke, Manton, and Clarke's series (1955–59) of studies of eight new species referred to this genus was the discovery that the third appendage is not a flagellum but a slender retractile filament with an adhesive tip, named the haptonema. In some species it may be extended to several times the length of the flagella and in all it evidently is used for temporary fixation to a substrate. It may be tightly coiled or fully extended during active swimming but it does not function as a locomotor organelle. For the first species described by Parke, Manton, and Clarke (1955, 1956), the electron micro-scope was used only to determine the external morphology of the whole body and appendages and the fine structure of the scales. Later (Parke, Manton, and Clarke, 1958, 1959; Manton and Leedale, 1961a, 1961b), thin sections of *C. chiton, C. strobilus, C. ericina, C. minor,* and *C. kappa* revealed the internal structure of the haptonema. The wall of this filament consists of three concentric membranes, rather widely separated in the fixed specimens. At least the outer membrane is three-ply. Within the innermost membrane, and often adherent to it, are six (*C. Strobilus* only), seven (all other species), or rarely eight (some individuals of *C. kappa;* with this exception the numbers are constant within the species) single tubular fibrils about 20–25 mμ in diameter. Unfortunately, few observations were made on the insertion of the haptonema in the cell; some fibrous structure, lying between the two kinetosomes, is apparent in *C. ericina* and *C. kappa.* The delicacy of the organelle and its tendency to rupture or blister during fixation (an especially frustrating problem in marine organisms) made it impossible to determine whether any matrix or other structural components were present in the haptonema, but many evidently intact specimens showed no other constituents. Since the whole filament is capable of tight coiling and rather rigid extension, it constitutes a particularly tantalizing example of a contractile organelle with fairly simple structure.

Two brief observations on whole mounts of chrysomonads by Pitelka and Dougherty (unpublished) may be mentioned in passing. *Monochrysis* is classed by Hollande (1952a) as a uniflagel-late, but in electron micrographs *M. lutheri* (Fig. 38, Pl. XI) is seen to have one long and one short flagellum, both simple, plus a tiny stub of a third appendage. A similar third stub exists in *Prymnesium parvum*, where the two flagella are more nearly equal in length. *Prymnesium* has been placed with *Chrysochromulina* in the triflagellate family, Prymnesiidae. Conceivably, the third appendage in these organisms might represent something com-parable to a haptonema; the existence of a third true flagellum in any chrysomonad remains to be demonstrated.

In the species of *Chrysochromulina* studied in thin section, the two or four chloroplasts are arranged parietally about the central nucleus, but distortions may occur with changes in body shape.

Stigmas are lacking. In *C. strobilus* and *C. ericina,* an elongate, spindle-shaped pyrenoid occurs centrally in each chloroplast. A very few lamellae or slender paired tubes, apparently continuous with the lamellae, traverse the pyrenoid. In *C. chiton, C. minor,* and *C. kappa* the pyrenoid occupies a diverticulum projecting from the chloroplast into the cytoplasm or, in *C. kappa,* into an invagination of the nucleus. In all species the pyrenoid contains moderately dense, granular or filamentous matrix material and adjacent regions of the chloroplast are of very low density, with few lamellae. The nature of the substance, if any, stored here is unknown; chrysomonads do not store starch. The characteristic photosynthetic product, leucosin, is identifiable by light microscopy in large vacuoles in the cytoplasm; these appear empty in electron micrographs since leucosin is extracted by dehydrating alcohols.

Membrane-bounded lipid droplets, food vacuoles, and microtubular mitochondria are scattered in the cytoplasm of *Chrysochromulina.* A surprisingly large and elaborate Golgi apparatus occurs anterior to the nucleus and immediately below the bases of the two flagella. The basal bodies lie at an angle to one another; in *C. strobilus* this is very obtuse, so that the kinetosomes themselves are nearly parallel to the cell surface and flagella emerge obliquely. No rhizoplasts or other root fibrils are observed in most species, but in *C. ericina* fibrous connections pass from lateral insertions on the kinetosome bases out to the cell membrane or under the membrane to adjacent chloroplasts.

In summary, a few morphological details may be noted that seem to be consistent for the chrysomonads thus far investigated. The chloroplasts generally are longitudinally oriented flanking the nucleus, with lamellae running parallel to their long axes. The two kinetosomes lie close but not parallel to one another. Free flagella emerge generally at the level of the cell surface, or from a relatively shallow depression. Rhizoplasts have been clearly seen in three genera, and flagellum-stigma associations are found in those species that posses stigmas. Golgi elements occur in a specific and constant location near the basal bodies of the flagella, but without evident physical linkage to these. Mitochondria are microtubular. Thus far, all chrysomonads studied in any detail have proved to be biflagellate, with the tentative exception of *Hydrurus.*

A few relevant observations on the flagellated reproductive cells of brown algae may be interposed here, since these cells show considerable resemblance to chrysomonads. The stigma-flagellum relationship in the spermatozoid of *Fucus* (Manton and Clarke, 1956) has been mentioned in Chapter 2. A particularly interesting situation exists in the spermatozoid of *Dictyota* (Manton, 1959b), where the uniflagellate cell conceals a second kinetosome, adjacent and nearly parallel to the flagellum-bearing one but ending blindly below the cell membrane. From the base of the flagellum-bearing kinetosome arises a fibrous rhizoplast which descends towards the nucleus, passes near its surface without making contact, then runs past one or more mitochondria that may be flattened against it, and finally ends just under the cell membrane. In *Fucus,* the two kinetosomes are nearly parallel but point in opposite directions; they seem to be in firm contact with each other and laterally with the nuclear surface. From the apposed surfaces of the kinetosomes a strand passes to the stigma and another toward the fibrous anterior proboscis; no rhizoplast is apparent. In all of these cells, mitochondria are microtubular.

Cryptomonadea, Dinoflagellatea

We come now to two well-defined phytoflagellate classes that often are considered to be related to each other but show no clear affinities with any other phytoflagellate groups. They usually are placed provisionally with the brown stock (Hutner and Provasoli, 1951), but their pigment complements are in some ways intermediate between these and the greens (Dougherty and Allen, 1959, 1960). Although the dinoflagellates are enormously abundant both in species and in individuals in marine and fresh waters, and as symbiotes in other organisms, they have not been sufficiently studied with the electron microscope to provide any conclusions of general significance. For the cryptomonads, a small class but one that includes some familiar laboratory species, we know even less; only their flagella and trichocysts have been examined.

At present writing, the author is aware of only one report concerning the ultrastructure of unquestionable dinoflagellate flagella. This resulted from a cursory examination of osmic-fixed whole cells of *Gyrodinium* sp. (Pitelka and Schooley, 1955). As is characteristic of dinoflagellates, this species has a long, ribbon-shaped transverse flagellum (typically held in a groove encircling the body) and a shorter, slender one directed posteriorly. The breadth of the transverse flagellum is explained by the presence of a unilaterally expanded sheath of low density. In some specimens a sparse array of long, flexuous mastigonemes was present, usually on one side only of the flagellum. The posterior flagellum was smooth and tapered to a short acroneme tip.

Observations on the nuclei of sectioned dinoflagellates (Grell and Wohlfarth-Bottermann, 1957; Grassé and Dragesco, 1957), with their enormous, helically wound interphase chromosomes, have been described in Chapter 2. The paper by Grell and Wohlfarth-Bottermann contains in addition a short account of other protoplasmic structures of *Amphidinium elegans*. The many small plastids are oriented approximately radially about the cell periphery and with the abundant starch grains fill up most of the extra-nuclear space. Lamellae run in orderly bands of usually three discs parallel to the long axis of the plastid. The cytoplasm contains many vesicular and tubular profiles and fine granules, microtubular mitochondria, and Golgi structures (noted but not illustrated). The pellicle consists of inner and outer membranes, both described as multilayered, separated by a rather wide clear zone that is traversed by irregular slender strands or tubules. Above the outer membrane is a fringe of projecting fine filaments, believed by the authors to represent the mucoid secretion of mucigenic bodies beneath the surface.

Like the scales of chrysomonads, the detached thecal plates of dinoflagellates have been examined intact in the electron microscope. Fott and Ludvik (1956b) found the plates of *Ceratium hirundinella* to be rather thick and homogeneous in texture, with evenly arranged depressions leading to circular pores.

Oxyrrhis marina is a small flagellate of uncertain position, classed as a dinoflagellate in the Grassé treatise (Chatton, 1952) but as a cryptomonad by some other authors. Dragesco (1952b) found no evidence of mastigonemes on its two flagella. Discharged trichocysts resembled those of some ciliates (p. 55).

Dragesco's observations (1951) on the trichocysts of the cryptomonad, *Chilomonas paramecium,* are cited on p. 55. The flagella of this species and its chlorophyll-bearing relative, *Cryptomonas* sp., were shown by Brown and Cox (1954) and Pitelka and Schooley (1955) to bear rather long, slender mastigonemes (Fig. 39, Pl. XI), pantomene in distribution but apparently less rigid than ochromonad mastigonemes. *Hemiselmis virescens* (Pitelka and Dougherty, unpublished) likewise has long, slender mastigonemes on both flagella.

Class *Phytomonadea*

The flagellated green algae, called Phytomonadea by zoologists and Volvocales by botanists, are of extraordinary interest to both disciplines because of their approach to multicellular organization. Clearly akin to the higher green algae and presumably to the land plants, they also have figured prominently in speculations on the origin of metazoa. Thus Haeckel's hypothetical blastaea was derivable from a hollow colony of flagellates resembling *Volvox* (Hyman, 1940), but modern heretics are engaged in vigorous dispute over metazoan origins (see Hanson, 1958; Greenberg, 1959). Relationship of the phytomonads to other specific groups among the protists remains obscure.

Credit for electron-microscope studies of the simplest green flagellates yet investigated is due once more to the energetic British botanists, Manton and Parke. Manton (1959a) made a detailed examination of a minute species until then known as *Chromulina pusilla* and assigned to a position in the Chrysomonadea, and came to the conclusion, based on pigment analyses as well as morphology, that it belonged more properly in or near the Phytomonadea. Manton and Parke (1960) subsequently redesignated the species as *Micromonas pusilla* and added a new species, *M. squamata* both to the genus and to their list of electron-microscope conquests. An additional small green flagellate, *Pedinomonas tuberculata,* was also studied.

Micromonas pusilla is the first phytoflagellate that upon electron-microscope scrutiny has clearly lived up, or down, to its reputation as a uniflagellate species. Except in cells believed to be dividing, sections contain a single basal body only; indeed there is hardly room for a second in the tiny cell, which averages 1 by 1·5 μ. The posteriorly-directed flagellum arises laterally from the pear-shaped cell body and consists, as noted earlier (p. 42), of a very short basal segment containing the conventional 11 fibers and a much longer terminal filament consisting of the two central fibers and the limiting membrane, with a small amount of matrix material. Whether this distal filament is actively motile or not has not been determined. The basal body of the flagellum lies immediately below the cell surface and approximately parallel to it; its distal end is marked by a transverse septum.

No rhizoplasts or other fibrous connectives associated with the basal bodies are detectable. The small nucleus, generally lateral to the kinetosome, contains a poorly-defined nucleolus and is surrounded by a double envelope.

The chloroplast, occupying a large part of the cell's volume, is roughly hemispherical, with a large eccentric pyrenoid surrounded by a thin shell of clear material believed (Manton and Parke, 1960) to contain starch. Between this shell and the plastid surface, lamellar discs in loose, discontinuous, sometimes overlapping stacks run circumferentially. This arrangement is rather different from the continuous longitudinal orientation of lamellae in chrysomonad chloroplasts, and first aroused Manton's suspicion of the accepted taxonomic position of the species. A subsequent pigment analysis revealed the presence of chlorophyll b, which is unknown in the Chrysophyceae but characteristic of green algae.

M. pusilla has a single, sausage-shaped mitochrondrion located near the anterior borders of chloroplast and nucleus, adjacent to the kinetosome. In Manton's preparations the internal mitochondrial membranes are sparse, but they resemble cristae rather than microtubules. A group of small rounded vesicles always seen close to the kinetosome is identified as the Golgi zone; a small fat droplet generally is found nearby. Other larger, empty vesicles occur here and there. The organelles pack the tiny cell so completely that the cytoplasmic matrix is reduced to a minimum. The cell is naked, limited only by one unit membrane. Micrographs of cells in division show that the basal body of the flagellum probably divides first, followed by the chloroplast, in which a cleavage furrow forms and progressively constricts the organelle through the pyrenoid, and then by the mitochondrion, which becomes U-shaped before fission.

As Manton points out (1959a, p. 329), "By the time that a cell is restricted to one nucleus, one plastid, one mitochondrion, one little group of golgi vesicles, one fat body, one flagellum, and a very small amount of residual cytoplasm, it seems doubtful whether any further reduction would be possible without involving the fundamental structure of the organelles themselves." The organelles have to be small to fit into a 1 μ body, but they appear no simpler in structure than their counterparts in other cells. The flagellum and kinetosome are, except for the shortness of the flagellum, of conventional dimensions. The sparseness of mitochondrial membranes and the apparent lack of flat cisternae in the presumed Golgi area cannot be taken as evidence of structural simplicity, since these features are known to vary with the metabolic states of cells. Whether *M. pusilla* is primitively or secondarily restricted to one of everything, it might provide an extraordinarily useful tool for the biochemist. It should be possible to obtain serial sections through an entire cell, and from them to calculate the actual total membrane area of the chloroplast and mitochondrion, for example. From estimated molecular dimensions, one could arrive at a figure approximating the number of macromolecules required to sustain the measurable metabolic activities of the cell.

M. squamata (Manton and Parke, 1960) is similar to *M. pusilla* in most respects, but in addition to its larger size (3–5 μ diameter) shows some interesting structural peculiarities. Like *M. pusilla,* it has a single large plastid with peripheral discs and a dense, eccentric pyrenoid surrounded by starch shells, a single mitochondrion, and a Golgi zone near the kinetosome of the single, posteriorly-directed flagellum. The mitochondrial membranes in the published micrographs appear to be microtubular, and the Golgi element is much more distinctly developed than it seemed in *M. pusilla*. The chloroplast contains a region of close-packed vesicular chambers identifiable as a stigma. Discontinuities in the chloroplast discs appear in surface view as rather uniformly distributed holes penetrating several discs. A single fibrous rhizoplast originates at the base of the kinetosome, passes through the cytoplasm parallel to the kinetosome and terminates at the adjacent nuclear membrane.

Of considerable interest is a coating of regularly overlapping, submicroscopic scales all over the surface of the body *and of the flagellum*. These are slightly mineralized, flat discs with a spider-

I

web marking of ridges and grooves. Manton and Parke observed occasional intracellular vesicles containing scales, and since phagotrophy was never observed in this species they concluded that scales are synthesized within the cell, perhaps in a prominent group of vacuoles with heterogeneous contents found near the mitochondrion. The regular imbrication of the superficial scales, particularly conspicuous on the flagellum, could be explained by the successive opening of vesicles, each containing one completed scale, at the body surface around the flagellum base.

In addition to scales, the flagellum of *M. squamata* often bears short, curved mastigonemes; these either are sporadic in occurrence or usually are detached from the flagellum in the process of preparation.

In *Pedinomonas tuberculata* (Manton and Parke, 1960), the cell membrane is elevated at intervals over cone-shaped tubercles of dense amorphous material; in addition, tiny flakes or spicules are scattered over the surface, but these are not scales, nor do they form a continuous wall. The living cell swims with its single flagellum extending from the posterior pole. In this position the single, large, curved chloroplast fills the anterior end of the cell. In electron micrographs the plastid is seen to contain a dense central pyrenoid and discontinuous peripheral discs converging at the two ends of the organelle. Starch grains surround the pyrenoid and appear elsewhere between discs. In the concavity of the plastid surface are several small microtubular mitochondria, and a conspicuous Golgi apparatus lies between these and the kinetosome. The nucleus is much flattened against one side of the body between the extremities of the chloroplast. Striated root fibrils arise from the proximal end of the kinetosome and pass beneath the cell membrane to unknown destinations.

The polarity of *P. tuberculata* and the two *Micromonas* species is of some interest. In saying that the flagellum arises antero-laterally in *M. squamata,* laterally in *M. pusilla,* or posteriorly in *P. tuberculata,* Manton and Parke are considering the end of the cell directed forward most commonly in locomotion to be anterior. However, the same cluster of organelles is found near the kinetosome wherever it occurs: Golgi body, one or more mitochondria, and nucleus, while the chloroplast occupies the opposite half or more of the cell's volume. Thus it is justifiable to say that the kinetosome complex marks the morphological anterior pole of the cell, and that *Pedinomonas* simply spends most of its time swimming backwards.

The flagellum of *P. tuberculata* bears a liberal garniture of extremely fine mastigonemes, more flexuous and much more delicate in appearance than those on the chrysomonad pantoneme flagellum, and quite dissimilar to the stout, short mastigonemes of *Micromonas squamata*.

Another green alga with distinct mastigonemes is the quadri-flagellate *Platymonas*. An unidentified species of this genus was studied briefly by Pitelka and Schooley (1955) and subsequently four strains of *P. subcordiformis* were surveyed by Pitelka and Lewin (Lewin, 1958). In all instances, all four flagella bear fairly abundant but (in the dried specimens seen) erratically distributed and oriented mastigonemes (Fig. 41, Pl. XI). Although phyto-monad affinities of *Platymonas* have not been seriously questioned, Lewin (1958) has shown that the cell wall contains polysaccharides, not cellulose, and has suggested that a critical analysis of pigment composition is to be desired. *Haematococcus pluvialis* (*Sphaerella lacustris*) in electron micrographs published by Mühlpfordt and Peters (1951) showed a thick, dense pile of very short fibrils surrounding the entire flagellum, even the tip. These did not resemble mastigoneme structures in any other flagellate and may have been artifacts. Pitelka (1949) reported smooth, simple flagella in the same species.

Other species of green algae studied with the electron micro-scope have proved to have simple or acroneme flagella. The spotty occurrence of mastigonemes in phytomonads and their frequently disorderly appearance (an artifact?) when they are present leaves wide open the question of their significance, both functional and phylogenetic.

The only other phytomonad flagellate genus for which published reports on ultrastructure are available is *Chlamydomonas*. *C. reinhardi* was examined in some detail by Sager and Palade (1957), who were primarily interested in chloroplast structure, and the flagellar apparatus of *C. moewusii* was studied by Gibbs, Lewin, and Philpott (1958). The *Chlamydomonas* cell (Text-fig. 7) is limited by a unit membrane and surrounded by a cellulose wall that appears in electron micrographs as a homogeneous or

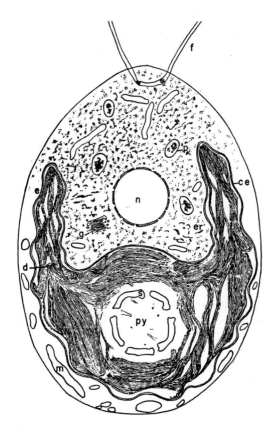

TEXT-FIGURE 7. Diagrammatic sketch of *Chlamydomonas*. ce, chloroplast envelope; d, lamellar discs of chloroplast — note discontinuous arrangement; e, eyespot granules within chloroplast; er, endoplasmic reticulum; f, flagellum; g, Golgi body, m, mitochondrion; n, nucleus; p, vacuoles containing metaphosphate; py, pyrenoid; s, starch plates. From Sager, 1959.

stratified zone of moderate density. At the anterior end of the cell is a small papilla from opposite sides of which the two flagella emerge to pass through tunnels in the cell wall. Within the cell, the two kinetosomes are attached to one another basally, so that they form a wide V when sectioned in their common plane. In *C. moewusii,* the distal end of each kinetosome is marked by a transverse diaphragm, but this appears to be traversed by a peculiar short cylinder, about 70 mμ in diameter by 120 mμ long, with a thin dense wall, occupying the center of the flagellum. The two central fibers arise immediately distal to the cylinder and do not appear to be attached to it. Projecting from the anterior faces of the fused kinetosomes is a dense material in the form of a cone-shaped chamber with a low-density lumen.

Gibbs, Lewin, and Philpott studied series of cells fixed at intervals after adhesion of mating pairs and found that within 10 minutes a bridge of homogeneous protoplasm connected the apical papillae of the two cells. A factor suppressing the motility of one of the mating partners is known to be passed through such a bridge. After one hour, a dense core appeared within the bridge, continuous with the anterior projections from the kineto-somes in each cell. Flagella of mutant strains of *Chlamydomonas* with impaired motility were studied by the same authors; except for decreased length in some paralyzed strains, the flagella all were normal morphologically.

Chloroplasts of *C. reinhardi* show the discontinuous stacks of lamellar discs described in Chapter 2. The central pyrenoid is traversed by tubules continuous with lamellar membranes and is surrounded by starch plates (Fig. 7, Pl. II). Starch appears first around the pyrenoid but may subsequently be deposited else-where in the plastid. Mitochondrial membranes resemble cristae more than tubules, but often seem to have the shape of irregularly oriented flat discs. Several compact stacks of Golgi lamellae with peripheral microvesicles occur near the nucleus. Granular membranes are well distributed through the cytoplasm, showing occasional continuity with the outer membrane of the nuclear envelope and with Golgi membranes. Larger empty vesicles with smooth membranes are seen, some possibly representing the contractile vacuoles but lacking a differentiated cortex.

Electron-microscope studies on other, non-flagellated green algae have been concerned mainly with chloroplast structure and provide little additional information relevant to our consideration of phytomonads. In general, chloroplasts resemble those of *Chlamydomonas,* Golgi bodies appear in the vicinity of the nucleus, and mitochondria have sparse, irregular cristae.

Class Euglenea

The euglenoid flagellates constitute a well-defined group resembling the phytomonads in the nature of their photosynthetic pigments, but differing from them in several morphological features and in the production of paramylum, a starch-like reserve material that is not stained by iodine. A whole spectrum of metabolic patterns ranging from obligate photoautotrophy to heterotrophy is illustrated by series of species found in nature, and apochlorotic strains also varying in the complexity of their organic needs are readily produced experimentally from species of *Euglena*. Thanks to such seductive biochemical qualities, combined with an interesting degree of morphological diversification and an ubiquitous distribution, they are among the most familiar protozoa. The two best-known genera, *Euglena* and *Peranema*, are the only ones to have been studied thoroughly in the electron microscope; information on flagellar structure of several more is available.

A flagellum of *Euglena gracilis*, prepared for electron microscopy by permitting unfixed or osmium-vapor-fixed cells to dry, is seen in Fig. 42, Pl. XII. The unilateral array of long (up to 4 μ), slender mastigonemes, and the striated bands bordering the flagellar axis, have been seen in species of *Euglena, Astasia, Phacus, Rhabdomonas, Entosiphon,* and *Peranema* (Brown, 1945; Pitelka, 1949; Chen, 1950a, 1950b; Houwink, 1951; Manton, 1952; Brown and Cox, 1954; Pitelka and Schooley, 1955). Like the pantoneme flagellum of the chrysomonad-xanthomonad-brown-alga group, the stichoneme flagellum of the euglenoids is a distinctive and characteristic type. However, Pitelka and Schooley (1955) found that under certain conditions — specifically when cells were fixed in osmium tetroxide buffered to a pH at or somewhat more alkaline than that of the culture fluid in which they were growing — the euglenoid mastigonemes remained smoothly and compactly wrapped around the flagellum (Fig. 43, Pl. XII). It has seemed unlikely that the thickset fringe of delicate, flexuous filaments could be anything but a hindrance to rapid movement of the euglenoid flagellum; but if these normally adhere to the flagellar surface they would, like the short, stout, presumably rigid mastigonemes of chrysomonads, effectively increase its diameter. The extraordinarily heavy anterior flagellum of *Peranema trichophorum* has a rather scant coating of mastigonemes, but the striated bands are particularly well developed. Pitelka and Schooley showed that these were present as two easily detached ribbons about 400 mμ wide and with a transverse striping regularly repeated at about 100 mμ. In Roth's (1959) micrographs of sectioned flagella they appear as two curved ribbons flanking the flagellar axis (Fig. 40, Pl. XI), each ribbon consisting of four to six layers of dense, striated material.

The following account of the ultrastructure of *Euglena* is derived from reports by Groupé (1947), Wolken and Palade (1953), Reger and Beams (1954), Wolken (1956, 1960), Roth (1958a), Pitelka and Schooley (1957 and unpublished), Ueda (1958), de Haller (1959, 1960), and Gibbs (1960, the latest and most detailed account). All of these authors used *E. gracilis* except de Haller, who studied *E. viridis*.

The euglenoid pellicle is marked by longitudinal spiral striations that may take the form of deep grooves and crests and may be elaborately sculptured. It contains no cellulose (Pigon, 1947). It may be extremely plastic in some species and quite rigid in others. At the anterior end of the cell is a conspicuous, flask-shaped chamber, the reservoir, within which the flagella arise and into which the contractile vacuoles discharge. The reservoir communicates with the exterior subapically via a canal of moderate width. Isolated pellicle fragments of *E. gracilis* examined by Pitelka and Schooley (1957 and unpublished) shed some light on the disposition and composition of the conspicuous striations. According to Pochmann (1953, 1957) the euglenoid cell has a fundamental bilateral symmetry, usually obscured by spiralization. Fig. 44, Pl. XII, showing an empty pellicle, supports Pochmann's conclusion. Lateral striae form two whorls which spiral in the same sense but would be symmetrical if the spirals were straightened out. Medial striae converge to enter the neck of the reservoir in close-set ranks. Pochmann has shown that during division the cleavage plane passes through the reservoir opening and down the middle of the cell, spiraling with the striae. How bilaterality is reestablished in the daughter individuals remains unexplained. At the posterior pole of the cell, the striae, greatly reduced in number by fusion, converge in a single whorl.

PLATE XII

FIG. 42. Part of dried whole flagellum of *Euglena gracilis* fixed at uncontrolled pH. × 18,000. From Pitelka and Schooley, 1955.

FIG. 43. Part of dried whole flagellum of *Euglena gracilis,* fixed at pH near that of culture medium. × 11,000. From Pitelka and Schooley, 1955.

FIG. 44. Dried empty pellicle of detergent-treated *Euglena gracilis*. The two poles are nearly superimposed. Upper arrow indicates whorl of fibers at posterior pole. Lower arrows show lateral whorls at anterior pole, on either side of neck of reservoir. × 3,600. From D. R. Pitelka and C. N. Schooley, unpublished.

FIG. 45. Small part of dried pellicle of detergent-treated *Euglena gracilis*. × 40,000. From D. R. Pitelka and C. N. Schooley, unpublished.

FIG. 46. Longitudinal section through anterior end of *Euglena gracilis,* showing longitudinal and radial fibrils around neck of reservoir. × 14,000. From Roth, 1958a.

FIG. 47. Section of *Euglena gracilis* nearly perpendicular to cell surface, showing hooked profiles of pellicular ridges, cross-sections of 2 or 3 tubular fibrils in cytoplasm at left side of each ridge. × 49,000. From S. P. Gibbs, unpublished.

PLATE XII

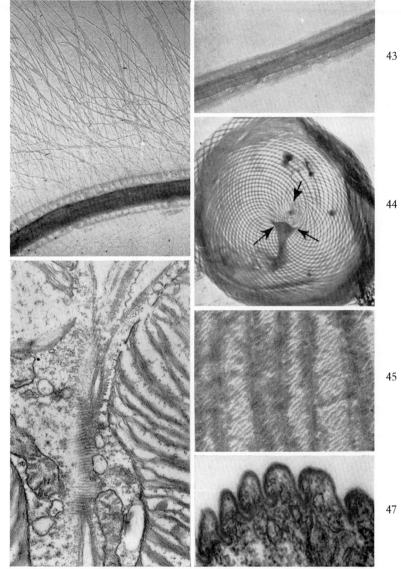

42

43

44

45

46

47

The structure of the pellicle is complex. Striae are evident in profile as strongly skewed ridges with an abrupt, even concave, slope on one side and a gradual one on the other. In fragments, it appears that a prominent fiber, about 30 mμ in diameter, runs along the peak of each ridge. The thin membranous band between fibers contains fine diagonal fibrils or folds, visible only on the outer surface of the pellicle (Fig. 45, Pl. XII). The pellicle most frequently tears along the bases of the grooves, suggesting that these are the weakest sites.

Examination of sectioned *E. gracilis* does not suffice yet to explain entirely the structure of the isolated pellicle strips. The same slanted or hooked profiles appear in cross-sectioned ridges. Running longitudinally within the overhanging peak of each ridge (Gibbs, 1960; de Haller, 1959, 1960) are two or three tubular fibrils, 18 to 25 mμ in diameter, which probably represent the continuous fibers seen in fragments (Fig. 47, Pl. XII). The diagonal markings on the outer pellicle surface are not explained. The surface seems to be limited by two membranes about 8 mμ apart; the outer is a typical three-ply membrane, the inner often is less distinct. At a level slightly below the bottoms of the grooves, a single layer of fine vesicles appears, slightly humped beneath the ridges.

In the reservoir, the inner one of the pellicle membranes seems to be absent. Fibrils similar to those in the pellicle crests but regularly spaced at shorter intervals run longitudinally in the neck of the reservoir and all around its bulb, immediately beneath the limiting membrane. In addition Wolken and Palade (1953) and Roth (1958a) found close-set fibrils situated deeper in the cyto-plasm, encircling the neck, but not the bulb, of the reservoir (Fig. 46, Pl. XII).

E. gracilis is commonly pictured as having a single flagellum with a bifurcated base inside the reservoir. Electron micrographs leave no doubt that two flagella are present basally, but no observations are reported to indicate whether or not the short one actually attaches at its tip to the long one. Both flagella arise from conventional kinetosomes, located just below the reservoir membrane. The basal 1 μ of each flagellum, above the kinetosome, is swollen and the central fibrils arise distal to this swelling. Higher in the reservoir, near the presumed point of junction of the two flagella, one of them bears the swelling known as the paraflagellar body (p. 24). The broadly expanded flagellar membrane here encloses a low-density space containing irregular patches of material, and a large (up to $1\cdot5\ \mu$), dense, ovoid mass. Beneath the adjacent wall of the reservoir is the cresent-shaped cluster of pigment-containing stigma granules.

From the bases of both kinetosomes, tubular fibrils pass out in various directions. At least some of these are continuous with the fibrils underlying the reservoir membrane and thus, presumably, with the fibrils of the pellicle system.

The disc-shaped chloroplasts of *E. gracilis* have regular, parallel bands of lamellae; each contains a pyrenoid consisting of a dense matrix traversed by thinner bands. Paramylum bodies typically are ranged on both surfaces of the pyrenoid regions of the plastid but outside its membrane; apparently they are not enclosed in membranes of their own. Like starch, they have a very low density in electron micrographs. The euglenid eyespot is unusual in being separated from any chloroplast; it is considered to be an entire modified plastid, but no observations on its ultrastructure during development have been reported.

The numerous, small, spheroid or rod-shaped mitochondria of *Euglena* have distinct cristae; often these project radially inward from the inner limiting membrane (see Fig. 4, Pl. I). Well-developed Golgi bodies, consisting of eight to 20 packed sacs with peripheral microvesicles, are distributed throughout the cell. The cytoplasmic matrix contains an abundance of Palade particles, and smooth endoplasmic reticulum. Conspicuous large vacuoles containing scattered vesiculate, very dense conglomerates are identified by Gibbs as metachromatic granules, or volutin. In *E. viridis* (de Haller, 1960), peripheral vesicles containing a dispersed coagulum of moderately dense material represent the mucigenic bodies, which are absent from *E. gracilis*.

The *Euglena* nucleus has a typical fenestrated double envelope with a highly irregular contour. The nucleoplasm is densely granular, with compact clumps interpreted by Ueda (1958) as sectioned chromonemata. The large persistent nucleolus or endosome is finely granular in texture, and denser than the nucleoplasm. Gibbs (1960) pictured an ovoid body consisting of three concentric rings of material resembling the nucleolar substance, lying in a concavity of the endosome.

Peranema trichophorum is a colorless biflagellate euglenoid that ingests particulate food through a permanent cytostome located next to the external opening of the reservoir canal. Since this is the first instance we have encountered, and one of the few known among the flagellates, of a specialized mouth structure, it is of particular interest. Its ultrastructure has been examined by Roth (1959). The living cell has a slender, rounded anterior end during normal swimming. In this position the cytostomal opening is small and obscure, but a pair of conspicuous, tapering rods, the pharyngeal rodorgan, lies next to the reservoir. The cell may feed on *Euglena* or other organisms as large as itself, either by ingesting them whole or by attacking them in small bites. During the latter process the pharyngeal rods are in vigorous movement, acting together apparently as a puncturing tool, probe, and lever (Chen, 1950a). The anterior end of the cell may be widely extended, permitting large ingesta to enter a food cup which always is separated from the smaller reservoir by the rodorgan (Pitelka, 1945). A system of what appear in stained specimens in the light microscope to be fibers articulates with the anterior ends of the rods; these surround the cytostome incompletely at rest and unfold during ingestion.

Roth's electron micrographs show that each pharyngeal rod (Fig. 48, Pl. XIII) is composed of a bundle of approximately 100 parallel, hexagonally packed tubular fibers, each about 26 mμ in diameter. Surrounding this bundle for most of its length is a dense, homogeneous zone and a heavy limiting membrane. The two rods are joined by a bridge at their anterior ends. The articulated cytostomal fibers actually are several sheet-like processes with a beaded or fibrillar structure. These insert in grooves on the pharyngeal rods at two different levels. Roth speculates that the sheets are contractile and thus are responsible for the movements of the rodorgan, which itself always moves as a rigid unit. This seems likely, and one would like to have more information on the detailed configuration of the sheets. To account for the energetic movements of the rodorgan, the sheets would have to attach somewhere to another structure with some tensile strength, and this probably could only be the pellicle. That such

PLATE XIII

FIG. 48. Cross-section of rodorgan of *Peranema trichophorum* showing fibrillar composition of two rods, and grooves into which sheets (seen here as dense strands) insert. × 36,000. From Roth, 1959.

FIG. 49. Section perpendicular to cell surface of *Peranema trichophorum*. Upper arrows point to fibrils underlying pellicle, lower arrow to flattened paraflagellar rod of trailing flagellum. × 50,000. From Roth, 1958a.

FIG. 50. Oblique section near anterior end of *Bodo saltans*. Trailing flagellum and part of its kinetosome at right. Kinetoplast is membrane-limited body enclosing filamentous reticulum and dense granules at top left. Under cell membrane near kinetosome is row of tubular fibrils. Part of nucleus at bottom right; food vacuole and numerous small vesicles in cytoplasm. × 45,000. From Pitelka, 1961b.

FIG. 51. Cross-section of flagellum of *Bodo saltans*, showing flagellar fibril bundle and paraflagellar rod in inflated sheath. × 56,000.

PLATE XIII

48

49

50

51

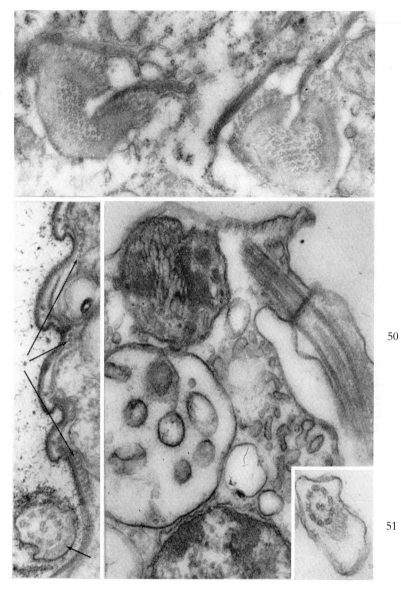

an insertion has not been seen by either light or electron microscopy is not surprising, considering the difficulty of tracing them at all. The important point is that, by all accounts, the movements of the rodorgan must be caused by something more precise than generalized protoplasmic contraction, and that in the structured sheets we perhaps have an intracellular muscle with a very specific action.

The leading flagellum of *Peranema* is assumed to be sensitive at least to mechanical stimuli (Lowndes, 1944; Chen, 1950a), and indeed it is difficult to escape this conclusion when studying the living organism. The second flagellum normally adheres to the body surface and does not participate visibly in swimming movements.

The *Peranema* flagellum, with its two ribbons of banded material and its coating of mastigonemes, was described above. In addition to these structures, Roth found a cylindrical rod of homogeneous material enclosed within the flagellar membrane. Near the bases of both flagella, the diameter of this paraflagellar rod exceeds that of the fibrous axis; cross-sections cut at this level resemble somewhat the flagellum of *Euglena* cut through its paraflagellar body. In *Peranema,* the rod extends through an unknown length of the leading flagellum outside the reservoir; it becomes flattened against the side of the trailing flagellum that is adherent to the body. The rods of both flagella may perhaps continue into the cytoplasm. A cross-section through the kinetosomes shows two additional ring-shaped profiles, somewhat smaller in diameter than the kinetosomes but like them composed of nine fibrils; Roth considers these to be the roots of the paraflagellar rods.

Many euglenoids show a type of body contortion confusingly known as metaboly. This generally appears to be a sort of ameboid movement restricted by the plastic but not notably elastic pellicle. The rather vigorous body movements of *Peranema* may perhaps involve something more than this. Roth believes that they do, and describes some structures associated with the pellicle that might be contractile. Like the surface of *Euglena,* that of *Peranema* is composed of at least two membranes, separated by space 3 to 4 mμ wide. A third layer, identified as a membrane, occupies the same position as the vesicular layer in *Euglena.*

Tubular fibrils, about 21 mμ in diameter and as many as eight or nine in number, run parallel to and beneath the grooves of the pellicle (Fig. 49, Pl. XIII). In addition, structures interpreted by Roth as heavy fibers are seen below the pellicular crests. Their diameter fluctuates between 130 and 230 mμ, and alternating regions of varying densities and vesicular enclosures indicate a periodic repeating pattern. They are present under only about half of the grooves seen in any one section, with no evident regularity in distribution. In view of the facts that very few longitudinal sections of these structures have been seen and that these show only short segments, their linear continuity is questionable. Certainly more information is required before any function can be attributed to them. No organelles seen are identified as the mucigenic bodies that occur in rows along the pellicular striae and are demonstrable by appropriate staining techniques in the light microscope.

To attempt a comparative summary of phytoflagellate ultrastructure based on the very small number of species sampled is obviously a precarious business, but it seems worth trying, if only to tempt future workers to demolish it with more information. Chloroplast structure probably is destined to be understood and categorized as soon as any. The simple, single-disc lamellae of the unicellular red algae and the segregation of discs into discontinuous piles in the phytomonads would seem to represent it extremes of development among the phytoprotists; evolution of grana in higher plants from the phytomonad type is probable, but morphological intermediates in this series have not been described.

It is tempting to ascribe some significance to the general occurrence of cristae in the mitochondria of green algae, as opposed to the microtubules found in all representatives of the brown stock (as well as the great majority of other protozoa). Both the land plants and the metazoa commonly have cristæ mitochondria, and for those who believe that metazoan organization is more likely to have stemmed from green algal colonie than from heterotrophic flagellates or ciliates, this may be welcome bit of evidence. But some microtubular mitochondria are found both in land plants and in animals (Rouiller, 1960 Novikoff, 1961), and we know little of the physiological meaning of variations in the configuration of mitochondrial membrane

The external morphology of flagella among the phytoflagellates is highly variable and would seem to be the result of opportunistic development in each group of mechanisms to increase efficiency. Internally, the morphology of the flagellar apparatus is fairly consistent throughout the phytoflagellates; the occurrence of accessory fibers leading from the kinetosome, in the form of either tubular filaments or striated roots, is widespread if not universal and foreshadows the much more complex development of such patterns in zooflagellates and ciliates. The existence of true rhizoplasts is confirmed in both phytomonads and chrysomonads, and their behavior in dividing cells needs to be analyzed. Polarization of the cell is conspicuously related to the location of the kinetosome in all flagellates, and extends to organelles with no obvious role in locomotor activities.

Phagotrophy occurs in probably all phytoflagellate groups, but no new information on the structures involved is available except for the highly specialized mouth of *Peranema*. Many phytoflagellates are markedly ameboid, some chrysomonads producing fine filose as well as lobose pseudopodia; no observations of these have been reported. Except in the euglenoids, cell membranes are uncomplicated.

CHAPTER 5

ZOOFLAGELLATES

THE CLASS Zooflagellatea as presently recognized is a polyphyletic assemblage of protozoa whose chief common characters are the presence of flagella in vegetative stages and the absence of a clear and direct relationship with phytoflagellate groups. Numbers of puzzling genera have in recent years been removed from the Zooflagellatea and more realistically classified as unpigmented derivatives of photosynthetic forms, and some others appear to be candidates for such a transfer. Many zooflagellates show pronounced ameboid tendencies, and as we have noted, very many rhizopods and actinopods have flagellated developmental stages, so that no natural boundary separates these large groups. Grassé (1952) divides the Class Zooflagellatea into three superorders: the Protomonadica, a heterogeneous assortment of small uni- and bi-flagellates; the Metamonadica, consisting of seven orders of multiflagellates; and the Opalinica, an enigmatic group of intestinal symbiotes of cold-blooded vertebrates, often considered to be allied to the ciliates.

Superorder Protomonadica
  The Order Choanoflagellida is a relatively small group of free-living species, characterized by the possession of a delicate protoplasmic collar or funnel around the base of the presumably single flagellum. They may attach to a substratum by a posterior peduncle, and often occupy a delicate hyaline or a silicified envelope. Several features suggesting a relationship with the chrysomonads are discussed by Hollande(1952b), who nonetheless feels that evidence is not sufficient to justify their classification with the phytoflagellates. Hollande, however, was not aware of a description by Lackey (1940) of *Stylochromonas minuta,* a chloroplast-containing chrysomonad with a protoplasmic collar surrounding the base of the single visible flagellum. Choano-

flagellates are of particular interest to zoologists because of their considerable structural similarity to the choanocyte cells of sponges, and are thus considered as possible representatives of a stem line leading to at least this phylum of animals.

The first electron-microscope study of a choanoflagellate was reported by Petersen and Hansen (1954) and dealt with whole mounts of *Codonosiga botrytis*. This species has a hyaline collar that is contractile and may be entirely withdrawn. In some electron micrographs a definite fibrous zone bordering the flagellum suggests the presence of poorly preserved mastigonemes of the ochromonad type; if correct this would constitute evidence of chrysomonad affinities. The collar is made up of fine individual thread-like processes.

Thin sections of the same species, studied by Fjerdingstad (1961b), show that the threads resemble orderly microvilli. They are 120 to 210 mμ in diameter, regularly spaced in a single ring; their cytoplasm appears undifferentiated and is continuous with that of the cell body. An amorphous material irregularly covering the limiting membrane of collar microvilli and cell body is interpreted as mucus. The author considers that the collar components are pseudopodial and retractable. In sections it is apparent that the cell is surrounded to a distance of some microns up the collar by a cup-shaped envelope composed of rather dense material with a suggestion of fibrillar constitution. Outside the collar, well below the rim of the envelope, is a deep invagination of the cell surface that Fjerdingstad considers to be the site of formation of food vacuoles. According to his interpretation, flagellar movements create a water current passing inward through the collar, which acts as a sieve to trap food particles that then are passed down its outer surface to the cell proper. The single flagellum and kinetosome of *Codonosiga* are of conventional structure. Fjerdingstad describes but does not illustrate striated fibrils at the bases of the collar and flagellum. Details of cytoplasmic structure are not discussed and it is not possible to determine from the published pictures whether the mitochondria, for example, are microtubular or cristal.

That the collar of the sponge choanocyte is essentially identical to the choanoflagellate collar is demonstrated in electron-microscope studies by Kilian (1954, whole mounts only), Rasmont

M

(1959), and Fjerdingstad (1961a). Fjerdingstad found that the neatly-spaced microvilli of the collar in *Spongilla lacustris* were connected by very fine lateral filaments to form a rather cohesive unit. Such filaments apparently are lacking in *Ephydatia fluviatilis*, examined by Rasmont, where the circlet of individual villi nevertheless maintains a rather impressive orderliness. The flagellum in both species is conventional, with an open, cylindrical kinetosome like those of protozoa. According to Rasmont, no root fibers are present. Additional studies of both the choanocyte and the choanoflagellate are needed to permit comparison of cytoplasmic structure and inclusions.

Grassé and Deflandre (1952) include in the Order Bicoecida a few genera of little-known organisms showing some resemblance both to the chrysomonads and to the choanoflagellates. They have two flagella, one of which is always directed posteriorly and attaches the cell to the bottom of its envelope, which may contain iron salts. This envelope has been examined (Robinow, 1956) in two species of *Bicoeca* (*Bikosoeca*), in which it resembles to a very striking degree the envelope of the chrysomonad *Stokesiella* (p. 112). Like the latter it is formed of a spirally wound membranous band. This is not surprising in view of the fact that species of the two genera were originally placed in the same genus. Unfortunately the flagella in Robinow's specimens were coiled within the envelope, so that no information is available on the presence or absence of mastigonemes.

If the choanoflagellates and bicoecids can be segregated as possible chrysomonad derivatives, the remaining three orders of the Superorder Protomonadica, while not revealing phytoflagellate affinities, do show relationship with each other and seem to constitute a natural grouping. The Bodonida include the only free-living species; the Trypanosomatida and the Proteromonadida are all symbiotic. Grassé (1952, pp. 47–49, 575) considers that the bodonids may represent an important stem group, leading through the proteromonads to the Metamonadica, and possibly implicated in the ancestry of the Sporozoa and even of the metazoa. As yet we have no information on the ultrastructure of the Proteromonadida, but various trypanosomes and two representative bodonids have been examined by electron microscopists.

Because of their importance as blood parasites of mammals, including man, the trypanosomatid flagellates have been subjects of a wide range of studies. The group includes internal symbiotes, pathogenic or not, of vertebrates, invertebrates, and plants; and a life cycle involving alternating insect and other animal or plant hosts is common. Different life cycle stages and different genera vary in body shape and size and in the location relative to the nucleus of the kinetosome and of an organelle known as the kinetoplast or kinetonucleus (see Grassé, 1952, or any textbook of protozoology or parasitology for characterization of genera and life cycles). They may be considered together here since our information for the most part concerns morphological features common to all of them.

Structure of the flagellum and pellicle in whole dried cells was reported by Kleinschmidt and Kinder (1950), Löfgren (1950), Das Gupta, Battacharya, and Sen Gupta (1951), Kraneveld, Houwink, and Keidel (1951), Das Gupta, Guha, and De (1954), Meyer and Porter (1954), Ray, Das Gupta, De, and Guha (1955), and Inoki, Nakanishi, Nakabayashi, and Ohno (1957). Sectioned material has been studied by the Inoki group (*op. cit.* and 1958), Anderson, Saxe, and Beams (1956), Chang (1956), Newton and Horne (1957), Horne and Newton (1958), Meyer, Oliveira Musacchio, and Andrade Mendonca (1958), Pyne (1958, 1960a), Pyne and Chakraborty (1958), Chakraborty and Das Gupta (1960), Clark and Wallace (1960), Meyer and Queiroga (1960), Newton (1960), Steinert (1960), and Steinert and Novikoff (1960).

In all species seen, the pellicle is conspicuously fibrillar (Fig. 53, Pl. XIV). Whole mounts show that the fibrils spiral about the body of the cell and converge, with reduction in number by fusion, at the two poles. In sections, the fibers are seen to be tubular, probably about 25 mμ in diameter (lower figures seem to represent measurements of one wall of an obliquely-cut fibril). They are evenly spaced in a single layer just below the continuous, three-ply, limiting membrane.

The flagellum has the usual 11 fibrils in a somewhat inflated sheath. In most published micrographs, a rod of dense material is evident lying close alongside the fibril bundle. In forms with an undulating membrane, the expanded flagellar membrane adheres firmly along one side to the body surface; the nature of

PLATE XIV

FIG. 52. Section through most of length of "cytostomal" tube of *Bodo saltans.* × 34,000. From Pitelka, 1961b.

FIG. 53. Whole dried cell of *Trypanosoma cruzi.* × 6,300. From D. R. Pitelka and W. Balamuth, unpublished.

FIG. 54. Part of undulating membrane of dried *Trichomonas* from termite gut. Partly detached trailing flagellum runs from lower left to right center. Edge of cell body at bottom. × 14,700.

FIG. 55. Section of *Tritrichomonas muris* showing part of axostyle with finely fibrillar wall at bottom center; edge of nucleus with encircling cytoplasmic membranes at left; part of parabasal body at top. × 50,000. From Anderson and Beams, 1959.

PLATE XIV

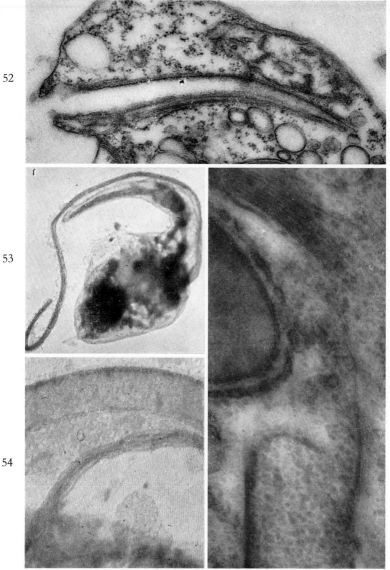

52

53

54

55

this adhesion is not demonstrated. The flagellum in all forms emerges from the cell at the base of a reservoir that is slenderer than, but otherwise like, that of euglenoids (Clark, 1959); a contractile vacuole empties into the reservoir.

The kinetosome typically projects for about half its length into the cavity of the reservoir; it is capped by a transverse diaphragm above which the central fibrils of the flagellar axis arise. Basally, the nine fibrils of the kinetosome end, in all forms studied, close to the kinetoplast; opinion differs as to whether or not actual contact with or even penetration of the kinetoplast membrane is demonstrable. The kinetoplast is a disc-shaped to ovoid body, usually situated with its long axis perpendicular to the kinetosome or slightly cupped below it. It is distinctly Feulgen-positive and has been shown by Steinert, Firket, and Steinert (1958) to incorporate tritiated thymidine. In addition, it reacts positively to mitochondrial stains. The function of this extranuclear, DNA-containing organelle is totally unknown in spite of ample and active interest in the subject. Natural occurrence of individuals lacking the kinetoplast has been observed, and akinetoplastic strains of some mammalian parasites may be maintained indefinitely *in vivo,* but not *in vitro* (Steinert, 1960). In the past, the kinetoplast has often been confused with a flagellar basal body or blepharoplast and with a parabasal body. It most definitely does not resemble either of these bodies morphologically, hence such misconceptions are no longer justifiable.

In its ultrastructure the kinetoplast is like no other familiar organelle (see Figs. 50, Pl. XIII; 5, Pl. I). It contains a reticulum of filaments showing a strong antero-posterior orientation, typically surrounded by a zone of varying density and limited externally by two unit membranes. The peripheral zone may be narrow or may occupy the bulk of a spheroidal kinetoplast, with the reticular structure constituting a plate or band across the middle. Membranous structures resembling mitochondrial cristae are seen in this peripheral region in most trypanosomes (Clark and Wallace, 1960), and Steinert (1960) was able to show in *Trypanosoma mega* a distinct mitochondrion completely continuous at one end with the kinetoplast. Small mitochondria also are scattered elsewhere in the cell. In all micrographs that resolve these clearly, their internal membranes have the form of cristae, irregular in orientation and rarely in contact with the limiting membrane.

In a site anterior to the kinetoplast and lateral to the reservoir, a Golgi element usually is seen. Since it does not appear to be attached to the kinetosome, it is a dictyosome rather than a true parabasal body. A larger vesicle in this region may represent the contractile vacuole, but it lacks distinctive cortical structure. The cytoplasmic matrix is fibrogranular and contains assorted profiles of endoplasmic reticulum. The nucleus has a perforate double envelope and a dense central nucleolus.

Two novel structures of particular interest have been reported: a basophilic, rod-shaped body in the cytoplasm of *Crithidia* (*Strigomonas*) *oncopelti* (Newton and Horne, 1957) and a fibrous tube tentatively identified as a cytostome in *Trypanosoma mega* (Steinert and Novikoff, 1960). Newton and Horne were able to isolate the rods, which always were located singly or in pairs in the posterior end of the cell. Analysis showed that they consisted largely of ribonucleoprotein, containing about 10 to 15 per cent of the total cell RNA. Apparently they reproduce by fission *in situ*. No evidence was available concerning their significance.

The "cytostome" demonstrated by Steinert and Novikoff in cultured crithidias of *T. mega* consists of a cylinder of fibrils apparently continuous with the pellicular fibril system, opening near the reservoir and curving deep into the cytoplasm; its internal terminus has not been identified. The cell membrane is depressed at the outer opening of the tube, forming a conical pit. Most of the length of the tube is thus filled with cytoplasm, but occasionally within it and frequently in linear arrays near it are found vesicles of about 180 mμ diameter, outlined by distinct unit membranes. By addition of ferritin to the culture medium, Steinert and Novikoff were able to show a concentration of ferritin molecules over the cell membrane at the open mouth of the tube, and also within the vesicles just described. The authors concluded that ingestion of ferritin by pinocytosis through the "cytostome" was occurring; confluence of ferritin-carrying vesicles formed larger inclusion bodies in the posterior end of the cell.

Although phagotrophy by trypanosomes has been suggested in the past, most investigators have remained skeptical; the organisms were assumed to survive by absorption of nutrients from blood and tissue cells. The common occurrence of micropinocytosis in vertebrate capillary epithelia and other cells that take up complex molecules makes the observation of the same phenomenon in blood parasites less than surprising. The existence of a highly specialized locus for this process is, however, unexpected. Although various intracellular fibers in trypanosomes have been reported by light microscopists, it is not certain whether any of these may correspond with the fibrous tube. If a similar structure exists in all other trypanosomes, or at all life-cycle stages, one would expect other investigators to have seen it in electron micrographs. Clark (personal communication) reports that he has encountered it very rarely in a series of species studied by him.

Pyne (1959, 1960b) has published two reports from a continuing investigation of *Cryptobia helicis,* a bodonid symbiotic in a garden snail, while the very common, free-living *Bodo saltans* is under study by Pitelka (1961b and unpublished). The two genera have many morphological features in common. Both have long trailing and shorter anterior flagella arising from a more or less pronounced anterolateral depression, near the base of which is an ovoid kinetoplast.

In both species, the flagellar membrane encloses, in addition to the 11 axial fibrils, a paraflagellar rod of dense material (Fig. 51, Pl. XIII) which in appropriate cross-sections is seen to be composed of six to eight flat bands about 7 mμ thick. This material originates immediately distal to the kinetosome and extends for an indeterminate distance along the flagellum — terminal sections lacking the rod are frequently encountered. The trailing flagellum of *Cryptobia* adheres in life to the whole length of the cell body. No physical connection is seen in electron micrographs, but the flagellum occupies a groove in the cell surface.

In both *Cryptobia* and *Bodo,* the kinetosomes project for a short distance into the circumflagellar depression. Basally, they stand in intimate contact with the limiting membrane of the kinetoplast. This body resembles the kinetoplast of the trypanosomes, but shows no pronounced orientation in the filamentous network occupying its center (Fig. 50, Pl. XIII). Peripherally, membranous profiles resembling mitochondrial cristae are present, and two unit membranes bound the organelle. In light-microscope preparations of *Bodo saltans,* a "cordon sidérophile" (Hollande,

1952c) arises from the kinetoplast, makes a complete longitudinal loop around the cell, and rejoins the kinetoplast. That this cord in all probability is a mitochondrion is indicated by its ultra-structure as shown in Figs. 1 and 5, Pl. I. The similarity to the situation demonstrated by Steinert in *Trypanosoma mega* is obvious, but the length of the single, continuous mitochondrion in *Bodo* seems to be unique. Pyne has not commented on mitochondria in *Cryptobia*.

Both species possess well-developed Golgi zones. In *Bodo* this occupies a constant position anterior and lateral to the kinetoplast and is often associated with large vacuoles. No fibrous connection to the kinetosome region is observed. Granular cytoplasmic membranes are seen in both species; in *Bodo* the cytoplasm is particularly rich in Palade particles. Food vacuoles containing bacteria are abundant; the contractile vacuole has not been distinguished from other, unidentified empty vacuoles.

The pellicle of *Cryptobia* is very similar to that of the trypano-somes; an outer unit membrane is underlain by a layer of parallel tubular fibrils, about 25 mμ in diameter. Fibrils in the wall of the flagellar depression appear thicker than those in the pellicle. Within the cytoplasm, two groups of tubular, 20 mμ fibrils are reported, leading away from the kinetosomes.

In *Bodo,* most of the body is limited by a unit membrane with no accompanying fibrils. But several sets of tubular fibrils, about 20 mμ in diameter, arise from the bases of the kinetosomes. Some of these directly connect the two kinetosomes; others run up the walls of the flagellar depression, immediately beneath the cell membrane, then curve around the asymmetric anterior tip of the cell and converge in an open apical cup. The latter corresponds to the rostral vacuole seen in the light microscope (Hollande, 1952c). From one side of this apical cup a group of fibrils descends into the cytoplasm, closely resembling the "cytostomal" tube of *Trypanosoma mega* described by Steinert and Novikoff (1960). In *Bodo,* however, the tube remains open, and lined by an invagination of the cell membrane, deep into the cell (Fig. 52, Pl. XIV). The cylinder of fibers terminates at the surface of the kinetoplast, adjacent to the kinetosomes of the flagella. In fact, the whole fibrous tube resembles a modified intracytoplasmic flagellum, lacking central fibrils and being instead lined by a membrane. No information on pinocytosis or phagocytosis has been obtained.

Similarities in the structure of the flagella, the kinetoplast, the pellicle, and the "cytostomal" tube where this occurs all suggest that the trypanosomes are closely related to the bodonids. Highly specialized physiologically for their symbiotic mode of life, they seem to have undergone minimal structural modifications. Of the various trypanosomes and the two bodonids investigated, the ubiquitous free-living *Bodo,* with its complex anterior fibrillar pattern, appears the least simple.

Certain similarities of bodonids to euglenoid flagellates may be mentioned: the pellicular fiber system, the accessory rod within the flagellar membrane as in *Peranema,* the reservoir (circumflagellar depression) with its system of fibrils arising from the kinetosomes. But it would be unwise at this stage to conclude that these resemblances are more than coincidental.

Superorder Metamonadica

The flagellates grouped by Grassé (1952) in the Superorder Metamonadica include those formerly placed in the orders Polymastigida, Trichomonadida, and Hypermastigida. The polymastigote and hypermastigote groupings clearly are artificial and Grassé, elevating several lower categories to ordinal rank, includes seven orders in the superorder. They are almost exclusively symbiotic, and several orders of them are known only from the guts of termites and wood-eating roaches. Unlike many animals of all phyla whose adaptations to symbiotic life have included general morphological simplifications and, particularly, reduction of locomotor organelles, these higher zooflagellates have indulged in a profligate elaboration of structure. A constellation of curious organelles, with additions and omissions, runs like a theme with variations through the whole assemblage. Prominent among these are the axostyle, a rod of varying thickness occupying an axial position in the body; and the parabasal apparatus, multiplied and embellished in a spectacular array of patterns. A characteristic set of flagella and associated organelles is known as a mastigont, and polymerization of mastigonts has led to series of increasingly elaborate morphological types, especially well demonstrated in the Order Trichomonadida

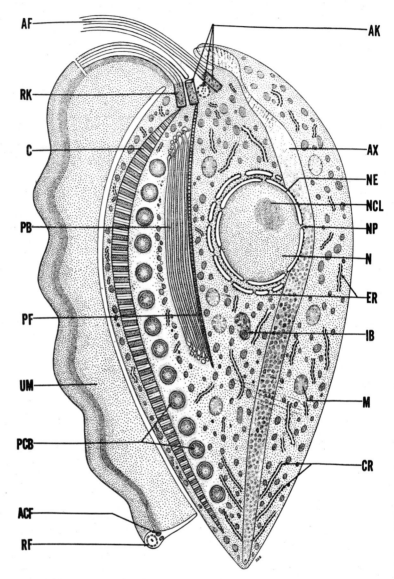

TEXT-FIGURE 8. Schematic drawing of *Tritrichomonas muris.*
AF, anterior flagellum; RK, kinetosome of recurrent flagellum;
C, costa; PB, parabasal body; PF, parabasal fibril, UM, undulating
membrane; PCB, paracostal bodies, ACF, accessory filament; RF,

(Kirby, 1949). Few other arrays of organisms provide such elegant material for the study of evolution of architectural complexity within cells.

Most metamonads have centrioles that appear in the light microscope to be distinct from the kinetosomes of the flagella; during cell division in many species the old flagella and their basal bodies, as well as other elements of the mastigonts, are dedifferentiated or topographically segregated well before new ones appear, and the latter then grow out from the centriole region. The centrioles are remarkable structures, often assuming an elongate form during part or all of the life cycle, with a proximal end duplicating to produce new centrioles and the distal end serving as the mitotic center. The extensive and meticulous studies of Cleveland and his colleagues (see Cleveland, 1957, 1960) on the extraordinary mitotic and sexual cycles of these organisms have provided new insight into the behavior of centrioles.

Ancestral trichomonad flagellates are considered to be central to the evolution of several other metamonad orders (Kirby, 1949; Grassé, 1952). Electron-microscope studies of species of *Trichomonas* (Ludvik, 1954; Inoki, Nakanishi, and Nakabayashi, 1959; Inoki, Ohno, Kondo, and Sakamoto, 1961) and *Tritrichomonas* (Anderson, 1955; Anderson and Beams, 1959, 1961) have been reported. These organisms (Text-fig. 8) bear three or four anterior free flagella and one recurrent flagellum attached by an undulating membrane to the cell surface. Below this attachment is an intracytoplasmic curved fiber called the costa. The axostyle is a hyaline rod that may protrude from the body at its pointed posterior tip; anteriorly it curves around one side of the central nucleus as a cup-like capitulum and extends to the apex of the cell. Just below this tip, a deep-staining zone called a centroblepharoplast by light microscopists is presumed to enclose a centriole and marks the point of convergence of all flagella, costa, and a parabasal filament to which the elongate, fusiform parabasal body adheres.

---

fibrils of recurrent flagellum; AK, kinetosomes of anterior flagella; AX, axostyle; NE, nuclear envelope; NCL, nucleolus; NP, nuclear pore; N, nucleus; ER, granular endoplasmic reticulum; IB, inclusion body; M, mitochondrion; CR, elements of the chromatic ring. From E. Anderson, unpublished.

FIG. 56. Section through anterior end of *Tritrichomonas muris*. All four kinetosomes are seen; that of recurrent flagellum continuous with base of flagellum at right. Parabasal fibril is opaque line attached to uppermost kinetosome; costa is broad, striated band attached to kinetosome of recurrent flagellum. Anterior end of axostyle is seen as row of sectioned tubules curving over kinetosome cluster. Undulating membrane, lower right, consists of expanded flagellar sheath enclosing fibrils and paraflagellar rod. × 60,000. From Anderson and Beams, 1961.

FIG. 57. Longitudinal section of anterior end of *Foaina* sp., showing four kinetosomes anterior to dense body identified as centrosome. Part of parabasal body at bottom. × 45,000. From Grassé, 1956b.

FIG. 58. Section through part of axostyle of *Joenia annectens*. × 45,000. From Grassé, 1956b.

Plate XV

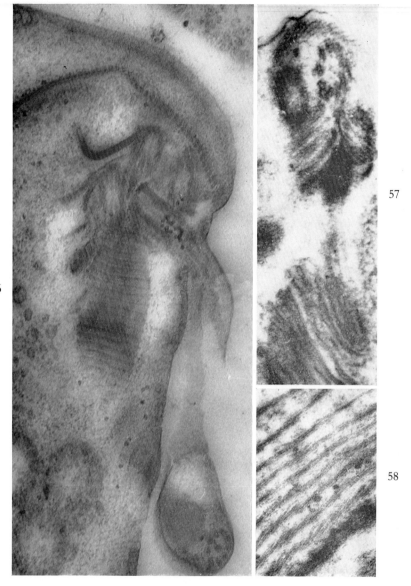

56

57

58

According to Anderson and Beams' accounts (1959, 1961), which provide the clearest micrographs of sectioned cells, the kinetosomes of all flagella are conventional hollow cylinders (Fig. 56, Pl. XV) formed by the nine fibrils that continue into the flagellum. Small dark granules occupy the centers of the kinetosomes, and a finely granular, rather dense matrix surrounds the whole cluster, but no separate identifiable centriole is detectable. That the matrix of the centroblepharoplast area has properties distinguishing it from the surrounding cytoplasm is apparent from its staining reactions and also from observations on fragmented cells. When termite-inhabiting *Trichomonas* is fragmented by ultrasonic bombardment (Pitelka, unpublished), a frequent component of the resulting brei is a mastigont lacking only the axostyle: flagella, costa, and parabasal filament are all present, diverging from a dense, bulbous central body.

The costa is a stout, striated rod that originates in contact with the kinetosome of the recurrent flagellum; its repeating pattern consists of subdivided light and dark bands at intervals usually under 60 mμ but varying in different regions of the same organelle. The parabasal filament is similar but much slenderer, and arises from the kinetosome of one of the anterior flagella. The Inoki group (1961) considers that the costa in *Trichomonas foetus* is a flat band at the cell surface, directly along the line of attachment of the undulating membrane. A heavy striated rod equivalent to the structure identified by other authors as the costa is also present.

Anderson and Beams found that the undulating membrane of *Tritrichomonas muris* consists of an expanded flagellar sheath enclosing, in addition to the fibrous flagellar axis, one or two dense granular strips believed to correspond to the "accessory filament" of light microscopy, and a finely fibrous sheet-like component. The site of adhesion of the membrane to the body surface is marked only by a homogeneous or finely filamentous substance of low density occupying a gap between the adjacent membranes. Perhaps the undulating membrane differs in structure in different species. Electron micrographs of whole cells of *Trichomonas* from a termite gut (Fig. 54, Pl. XIV) show a broad undulating membrane consisting of a finely fibrillar sheet bordered by a homogeneous band (Pitelka, unpublished). The recurrent flagellum normally follows the inner edge of this band, but in the dried specimens may become detached. Grimstone (1961) has published a micrograph of a section of termite *Trichomonas* showing a distinct fold of the cell surface, with the flagellum, surrounded by its own membrane, adhering to one side of the fold. The flagellar membrane is greatly expanded and encloses what appear to be several sheets of fine filaments. It appears that the undulating membrane of the termite *Trichomonas* includes a true membrane contributed by the cell surface, while that of *Tritrichomonas* is merely an inflated and reinforced flagellum without intimate connection to the cell.

The axostyle of *Tritrichomonas* is a tubular structure, limited by a sheet of fine (14 mμ) fibrils and filled with spheroidal granules of varying size and moderate density (Fig. 55, Pl. XIV). Anteriorly, the surface of the axostyle facing the nucleus appears to be interrupted, so that its cavity is open to the cytoplasm. At the cell apex, the tip of the organelle extends around the kinetosomal complex without making any direct contact with it. At the posterior end of the cell, the structure identified by light microscopists as a chromatic ring encircling the axostyle is seen as a cluster of paired granular membranes. Where it protrudes from the posterior end of the cell, the tip of the axostyle is covered by an extension of the cell membrane.

The cytoplasm of *Tritrichomonas* contains an abundance of Palade granules, scattered granular membranes, and numerous very small vesicles. In addition, ovoid bodies called chromatic granules by light microscopists are distributed through the cell but are particularly conspicuous aligned alongside the costa. These are limited by double membranes and contain a rather densely granular material. Other bodies of similar size are identified by Anderson and Beams as mitochondria; their internal membranes are indistinct. Inoki and his colleagues (1959) were unable to find mitochondria in *Trichomonas vaginalis*. The nucleus in Anderson and Beams' micrographs is rather uniformly dense, with more compact zones constituting the nucleoli. Surrounding the annulate double envelope is a layer of circumferentially oriented granular membranes.

A single micrograph of the trichomonad flagellate *Foaina* (Grassé, 1956b) shows a number of structures of interest (Fig. 57, Pl. XV). Four kinetosomes lie close together, and just behind them a very dense, homogeneous body identified as the centrosome. The anterior tip of the axostyle, as in *Tritrichomonas,* curves over the top of the kinetosome cluster, but does not in this picture make contact with the centrosome. The axostyle is cut tangentially and resembles the wall of the *Tritrichomonas* axostyle. A striated fiber passes from the centrosome posteriad to the parabasal body.

A detailed and beautifully illustrated study of *Pyrsonympha vertens* was published in 1956 by Grassé. This termite-inhabiting flagellate is representative of a group whose affinities with the trichomonad line are not clear. As seen in the light microscope, the cell has the shape of an elongate pear, with four or eight recurrent flagella inserting on a centrosome at the anterior end and spiralling about the body surface to become free beyond its posterior pole. From the centrosome also arise a long, slightly flexuous axostyle and a shorter paraxostyle. At some periods in the life cycle an organelle of attachment extends from the cell's anterior tip into the tissues of the host gut.

Grassé's electron-microscope studies show that the flagella rest in well-marked shallow grooves in the cell surface (Fig. 59, Pl. XVI), from which they may occasionally be dislodged by mechanical disturbance. The apposed membranes of cell and flagellum are parallel and separated by a fairly constant gap. Laterally, the flagellar membrane is expanded and convoluted and often appears to engage like the teeth of a gear with crenellated ridges on the cell surface bordering the groove. The flagellum itself is of peculiar structure. The two central fibrils sometimes — but not always — seem to be accompanied by one or two supplementary ones. Besides these and the usual nine peripheral fibrils there are from two to five additional linear structures. These are dense and triangular in cross section, with the apex of each triangle in contact with one of the peripheral fibrils of the flagellar axis on the side facing away from the cell. The dense material may appear homogeneous but more commonly encloses up to five circular profiles resembling single flagellar fibrils. Distinct but much smaller densities often appear on the medial side of the axial fibril bundle.

In spite of the fact that they adhere continuously to the cell surface, the flagella of *Pyrsonympha* are motile; according to

PLATE XVI

FIG. 59. Cross-section of flagellum of *Pyrsonympha vertens* lying in groove on body surface. Note accessory material between outer flagellar fibrils and membrane. × 50,000. From Grassé, 1956b.

FIG. 60. Transverse section of axostyle of *Pyrsonympha vertens*. × 45,000. From Grassé, 1956b.

FIG. 61. Longitudinal section of anterior end of *Pyrsonympha vertens*. Rounded, dense body at apex is centrosome. Diverging from it are seen, moving clockwise, dense paraxostyle, striated axial ribbon, and one kinetosome. Conical tip of nucleus at bottom. × 50,000. From Grassé, 1956b.

FIG. 62. Transverse section of *Lophomonas striata* at level of nucleus, showing fibrillar nature of calyx bands. Edge of nucleus at left. × 40,000. From Beams, King, Tahmisian and Devine, 1960.

Plate XVI

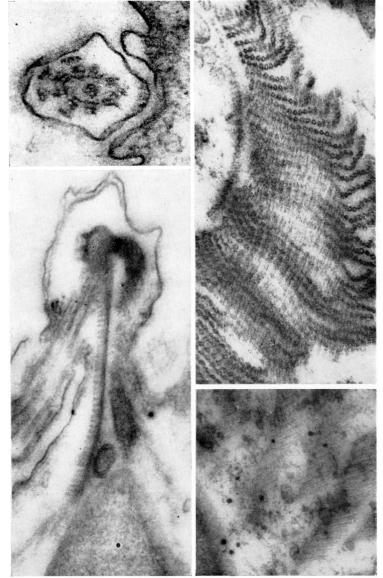

Grassé, waves of contraction, necessarily involving the adjacent cell surfaces as well, may be observed to pass down them. This sort of activity must pose mechanical problems not met by free-swinging flagella. Conceivably the dense strips on the outer half of the flagellum might offer mechanical resistance, balancing the drag of the body surface on the opposite side. They are rather similar to some structures seen in animal sperm tails (see Fawcett, 1958).

Within the cell, the kinetosomes apparently all originate in contact with a homogeneous, rounded body of moderate density, identified as the centrosome (Fig. 61, Pl. XVI). The latter is near the extreme anterior tip of the organism, and all kinetosomes are directed posteriorly.

The axostyle of *Pyrsonympha,* unlike that of trichomonads, is a contractile organelle; according to Grassé its movements are clearly distinct from, and not synchronous with, those of the flagella on the body surface. Its ultrastructure is also quite different from what we have seen in *Tritrichomonas* and *Foaina*. It is composed of a pile of longitudinally oriented lamellae (Fig. 60, Pl. XVI). In cross-section, anywhere from 14 to 74 more or less concentrically curved sheets may be counted in the pile, each one consisting of a single row of tubular fibrils (measured from Grassé's published micrographs at about 23 mμ in diameter). Lateral connectives are present between fibrils in a sheet, but none between adjacent sheets. One sheet of fibrils, instead of parallel-ing the others, is always seen perpendicular to them along one side of the organelle; in some specimens this may curve around, partly encircling the pile. Anteriorly, the axostyle ends a micron or so short of the centrosome, but its anterior tip is closely applied to a finely striated fiber or axial ribbon that inserts on the centrosome. The structure of this ribbon is somewhat enigmatic; its striations appear longitudinal where it is in contact with the axostyle, but closer to the centrosome it has a dark fiber along one side and transverse striations along the other. One of Grassé's micrographs, of a cell at a stage when the nucleus occupies an anterior position (Fig. 61, Pl. XVI) shows the same axial ribbon also in close apposition with the pointed upper end of the nucleus.

The paraxostyle is seen as a sinuous rod of very dense, perhaps filamentous material, inserting on a lateral face of the centrosome

N

(Fig. 61, Pl. XVI). The attachment organelle is a cigar-shaped projection, with a membrane underlain by spirally twisted fibrils enclosing a granular cytoplasm. At its base is a row of dense, fibrous structures known as the basal plaque, and, below this, a root that is indistinguishable in structure from the axostyle. The position of this root relative to other anterior organelles is not described.

One of the more remarkable features of *Pyrsonympha* is its apparent lack of any parabasal apparatus. No structure resembling a Golgi element was encountered by Grassé, who is a past master at discovering Golgi bodies. Other pyrsonymphids are believed to possess them, but members of one related order apparently do not. Grassé also was unable to find any organelles identifiable as mitochondria in *Pyrsonympha*.

In a complex metamonad, *Joenia*, mentioned briefly by Grassé (1956b), a striated axial ribbon connects the centrosome to the noncontractile axostyle, which is formed of concentric layers of fibrils (Fig. 58, Pl. XV) rather resembling those in the axostyle wall in *Tritrichomonas* and in *Foaina*. Grassé states that an axial ribbon exists in trichomonads he has examined, and believes that all of these axostylar connectives are homologous centrosomal derivatives, about which the inert or contractile fibrils of the axostyle differentiate secondarily. He does not comment on the resemblance of the axial ribbon to the rhizoplasts of some phytoflagellates and to the parabasal filament in trichomonads.

Among the higher metamonads (hypermastigotes) that are thought to be derived from trichomonads, genera representing three different orders have been examined with the electron microscope. Beams, King, Tahmisian, and Devine (1960) have studied *Lophomonas striata* (Order Lophomonadida) from the cockroach gut, and Beams, Tahmisian, Anderson, and Wright (1961) report on *L. blattarum*. Pitelka and Schooley (1958) and Grimstone (1959a, 1959c, 1961) have examined species of *Trichonympha* (Order Trichonymphida) from termites, and Gibbons and Grimstone (1960) have reported on details of flagellar structure in *Trichonympha*, *Pseudotrichonympha* (Order Trichonymphida) and *Holomastigotoides* (Order Spirotrichonymphida of Grassé) from termites or the woodroach. All of these organisms are uninucleate but bear large numbers of flagella and, typically, of some other elements of the mastigont.

Of the species studied, the least outrageously complex in its morphology is *Lophomonas striata* (Text-fig. 9), a rather small, spindle-shaped cell with a tuft of 100 or more flagella arising from a round plate slightly recessed into the anterior pole. From the

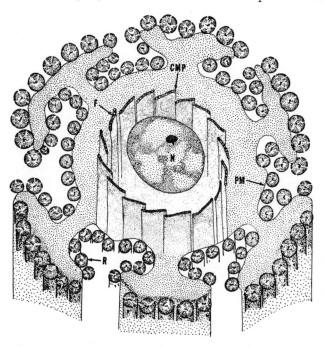

TEXT-FIGURE 9. Schematic drawing of a transverse section of *Lophomonas striata*. CMP, discontinuous bands or plates of the calyx; F, fibrils composing the calyx bands; N, nucleus; PM, cell membrane over the convoluted folds of the surface; R, rods adhering to the surface. From Beams, King, Tahmisian and Devine, 1960.

circumference of the plate, a cylinder called the calyx, consisting of what some light microscopists believe to be multiple axostylar fibers (Kirby, 1949; Grassé, 1952), descends posteriad, enclosing the axial nucleus and tapering to an end in the postnuclear cytoplasm. Peculiar, longitudinally oriented, rod-like striations have been variously interpreted as being cytoplasmic bodies or superficial symbiotic organisms. The micrographs by Beams and colleagues show that the flagella arise from kinetosomes equally spaced in concentric, incomplete circles. The basal bodies are somewhat elongate (about $0·75 \mu$) but of conventional structure, with a suggestion of a transverse septum at the level of the cell membrane, above which the central flagellar fibril pair appears. A rather ill-defined fibrous material interconnects the basal bodies at two different levels, one at their bases and one about three-fourths of their length distally. In addition, a network of fibers of unknown pattern occurs in the cytoplasm between basal bodies and nucleus. The origin of the calyx and its relationship, if any, with the basal bodies are not explained.

Apparently the calyx extends around the tuft of flagella for at least a short distance above the recessed plate of basal bodies. Here it is seen immediately below the unit cell membrane, as a uniform fibrous or striated layer. Below the proximal ends of the kinetosomes, it separates into bands (presumably the axostylar fibers of light micrographs). Transverse sections of the cell in the region of the nucleus (Fig. 62, Pl. XVI) show the calyx components as skewed, overlapping ribbons with a finely fibrillar structure resembling the axostyle wall of *Tritrichomonas, Foaina,* and *Joenia;* gaps between them permit continuity of the cytoplasm internal and external to the calyx. The termination of the bands posteriorly is not known. The nucleus of *Lophomonas* is small, spheroid, and remarkably homogeneous in the material studied. Cytoplasmic structures include miscellaneous small vesicles and tubules, fairly abundant Palade particles, Golgi elements, and bodies identified by the authors as mitochondria, although no internal membranes are evident in the published reproductions.

The entire surface of *L. striata* is deeply folded; the folds generally run anteroposteriorly and are convoluted and subdivided. Clinging to the surface everywhere are longitudinally oriented rods that show sufficient resemblance to bacteria that identification of the characteristic striations of this species as adherent microorganisms seems certain.

L. blattarum resembles *L. striata* in general morphology but lacks the surface convolutions and adherent rods. Its calyx bands converge to form a long axial rod posterior to the nucleus. A peculiar body surrounding the nucleus and previously designated as a parabasal apparatus is shown by Beams and colleagues to consist of an extraordinary system of radiating tubules. According to the authors' reconstruction, these originate at or possibly within the nucleus, extend outward, with some anastomoses, through the calyx, become highly convoluted, and end in enlarged bulbs filled with dense material. Their significance is unknown and they resemble no other known organelle, least of all a conventional parabasal or Golgi apparatus. Golgi bodies (apparently without parabasal filaments) and mitochondria are described in the cytoplasm of *L. blattarum* outside the calyx.

The formidable complexity of the trichonymphid flagellates almost defies description in a limited space. *Trichonympha* is typically bell-shaped, with thousands of flagella arising in longitudinal rows on the anterior part of the body. At the anterior end, a hyaline apical cap surmounts the rigid, flagellated rostrum, which is partly separated from the remaining flagellum-bearing zone by a deep circular fissure. Pitelka and Schooley (1958) were able to describe the chief architectural features of the flagellated portion of the cell, and their results are summarized in Text-fig. 10. The body is limited by a unit membrane which, on the apical cap only, is underlain by three parallel sheets of fine fibrils. The axis of the rostrum is occupied by an elongate, cone-shaped rostral tube, whose wall consists of radially arranged, striated, longitudinal ribbons enclosed in a membrane (Figs. 63, 65, Pl. XVII). At the flaring posterior end of the tube, this membrane disappears and the freed bands become slightly sinuous and pass posteriad to make contact with the cord-shaped parabasal bodies clustered about the nucleus. The rostral tube thus is composed of packed, striated parabasal fibrils similar to the single parabasal fibrils of trichomonads. Their striations have a period ranging from about 30 mμ in the rostral tube to 60 mμ posteriorly.

It is characteristic of the trichonymphids that the flagella and associated structures are divided into two (or four) symmetrically placed sets that separate preceding cell division and are retained in the daughter cells, which elaborate new sets to make up the adult complement. This bilaterality is apparent in cross-sections of the *Trichonympha* rostrum, where the rostral tube wall is seen as two opposed crescents, with a delicate striated or fibrillar lamella passing radially from one tip of each crescent. At the

PLATE XVII

FIG. 63. Approximately transverse section of rostrum of *Trichonympha* sp., showing rostral tube and radiating rows of kinetosomes. × 11,500.

FIG. 64. Section from centriole region of *Barbulanympha* sp. Part of centriolar complex appears finely filamentous. Arrows indicate two kinetosomes that are distinctly out of alignment with those of flagellated areas. × 28,000. From J. E. Cook, unpublished.

FIG. 65. Oblique thick section through anterior end of rostrum of *Trichonympha* sp., showing hemispherical complex at top center, parabasal filaments composing rostral tube wall, posterior extension of centriole in lumen of tube, rows of radiating kinetosomes. × 15,000. From Pitelka and Schooley, 1958.

FIG. 66. Section through posterior extremity of centriole in dividing *Barbulanympha* sp., showing spindle fibers diverging from homogeneously dense centrosome. × 3,500. From J. E. Cook, unpublished.

PLATE XVII

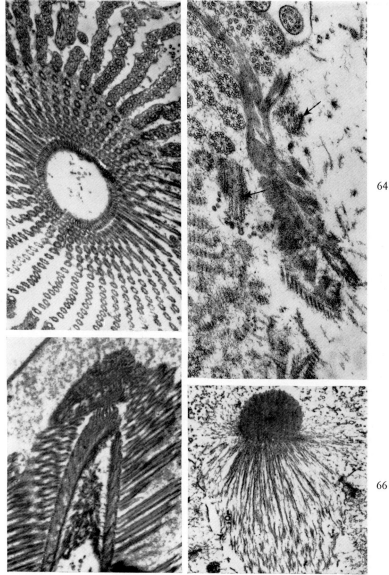

63

64

65

66

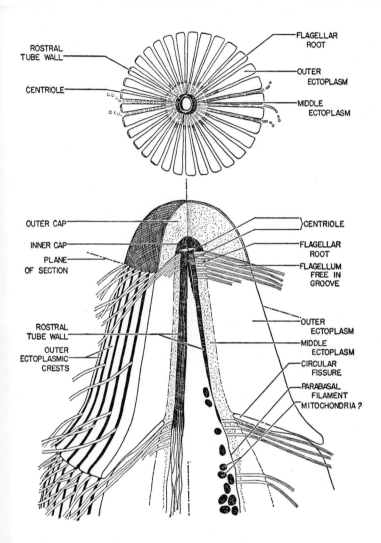

TEXT-FIGURE 10. Schematic drawings of the anterior end of *Trichonympha*. Below: body surface shown at left, cut away to reveal internal structures at right. Above: section representing plane indicated by oblique broken line on lower figure. From Pitelka and Schooley, 1958.

anterior end of the rostrum the parabasal fibrils appear to insert on a pair of flat, stratified half-discs that are partly embedded in a hemisphere of dense granular material (Fig. 65, Pl. XVII). Extending posteriad from this hemispherical complex, a rod of similar dense material occupies the center of the rostral tube for a short distance. According to Cleveland's (1960) light-microscope studies, the hemispherical complex represents the anterior ends of the two centrioles and the rod the posterior extension of one of them; at division the second also elongates and the mitotic apparatus will be formed between the two posterior ends. Some evidence of oriented lamellar structure is suggested in the central rod in the micrographs of Pitelka and Schooley, but observations of this critical area were too few to be conclusive. Grimstone (1961), however, reports from unpublished studies by himself and Gibbons the finding of a fibrillar centriole embedded in the hemispherical complex. In the related trichonymphid flagellate *Barbulanympha,* currently under study by J. Cook (personal communication), neat rows of kinetosomes diverge from a filamentous mass that represents part of the centriolar complex. Associated with it (Fig. 64, Pl. XVII) are two kinetosomes that are markedly out of alignment with the others. The posterior end of the centriolar complex in the same species, sectioned during mitosis, is a homogeneously dense, round body from which radiate the tubular spindle fibrils (Fig. 66, Pl. XVII).

Kinetosomes of the rostral flagella in *Trichonympha* stand in intimate contact with the outer surface of the rostral tube; posterior to the circular fissure they stand free in the ectoplasm. The parabasal fibrils twist about in the near vicinity of postrostral kinetosomes, but there are fewer fibrils than postrostral rows of flagella, and in any event the parabasal fibrils leave the rows of kinetosomes when they curve medially to meet the parabasal bodies. Kinetosomes, particularly on the rostrum, are extremely long — up to 4 or 5 μ. Their nine component fibrils continue as the peripheral fibrils of the flagella, which emerge from the cytoplasm at the bases of deep longitudinal grooves.

Considerably more detail concerning some aspects of these flagellar structures is provided by the beautiful micrographs of Gibbons and Grimstone (1960) The structure and orientation of the flagellar and kinetosomal fibrils have been described in

Chapter 2, but their orderliness can only be appreciated by reference to the original work. Flagellar morphology is essentially similar in *Trichonympha, Pseudotrichonympha,* and *Holomastigotoides.* In the first two genera, a cartwheel structure consisting of a central cylinder and radiating spokes occupies the proximal half of the kinetosomes. In *Trichonympha* the distal portion of the kinetosome is empty but in *Pseudotrichonympha* it contains a cluster of three delicate cylinders in kinetosomes on some parts of the body and a mass of dense granules in others. In *Holomastigotoides* the whole kinetosome appears empty. Very sinuous fibrous ribbons lie close to the postrostral basal bodies in *Pseudotrichonympha,* somewhat like the parabasal filaments of *Trichonympha.* In *Holomastigotoides* a regularly undulating ribbon of fine filaments with periodic transverse densities makes intimate connection with the kinetosomes in each spiral row near the posterior end of the body. Anteriorly, a thick, homogeneous band lies next to the basal bodies; nearby run ribbons of tubular fibrils that Gibbons and Grimstone identify as axostylar fibers.

Grimstone's (1959c) observations on the cytoplasmic and nuclear membranes of *Trichonympha* have also been summarized in Chapter 2. His picture of the parabasal fibrils shows that the repeating period, 36 to 54mμ in length, contains secondary bands within it. Publication of Grimstone's more recent morphological and cytochemical work on *Trichonympha* will undoubtedly clear up many additional questions left unanswered by the studies of Pitelka and Schooley. One of these is the nature of ovoid, membrane-bounded bodies with very fine canalicular contents provisionally identified by the latter authors as mitochondria.

Reflecting on the morphology of the flagellar apparatus and associated structures in the flagellates as a whole, one cannot fail to be impressed by the versatility of the cytoplasmic locus morphologically identified by the presence of the kinetosome. At the same time, the repetition of similar patterns in very dissimilar flagellates emphasizes the underlying analogy — and surely homology — of the apparatus wherever it is found. The constancy of flagellar and kinetosomal fibril pattern needs no further comment. Superficial elaborations such as mastigonemes and scales are restricted to the phytoflagellates and have been discussed in the preceding chapter. Accessory structures enclosed within the flagellar membrane occur in *Peranema*, in the localized paraflagellar body of *Euglena*, in the flagella of trypanosomes and bodonids, in the recurrent flagellum of some trichomonads, and in the adherent flagella of *Pyrsonympha*. Except for the paraflagellar body, these are lengthy rods or bands of dense or fibrous material. Their distribution suggests that they may function as stiffening or supporting structures, as proposed above for *Pyrsonympha*. They usually appear in flagella that adhere to the cell body, but are present in some free flagella as well. Critical studies of flagellar movement or sensitivity in series of species with variously modified and with simple flagella have not been attempted but certainly are needed.

Very little can be said about observed differences in the narrow region at the base of the flagellum where the central fibrils arise and where the cell membrane establishes its relationship with the flagellum. This must be in all flagella a particularly critical region both physiologically and developmentally; the more detailed our observations of its ultrastructure and variations, the more abysmal seems our ignorance of what actually goes on here. We face a similar lack of food for speculation as regards the structures seen occupying the center of the kinetosome in several species, as well as the fine fibrillar interconnections among kinetosomal fibrils and subfibrils illustrated by Gibbons and Grimstone. Their finding of elaborately different kinetosomal contents within the same species indicates that the matter must be of considerable importance to the kinetosome, but we cannot begin to guess why or how.

At least two fairly distinct types of fibrils arising from kinetosomes or centrioles have been widely observed. The first of these includes the striated rhizoplasts and other striated root fibrils of some chrysomonads and phytomonads, the striated parabasal fibrils of trichomonads, *Foaina*, and *Trichonympha*, the costa of trichomonads, and the striated axial ribbons of *Pyrsonympha* and *Joenia*. They vary widely in diameter; their repeating pattern is from 30 to 60 mμ in length. In addition to their periodic structure and kinetosomal origin, they have other features that recur repeatedly. Intimate association of their distal regions with the nucleus is seen in several phytoflagellates and again in *Pyrsonympha*, while it is known that in many trichonymphids the parabasal bodies (which have their fibril attachments on their nuclear faces) are applied closely against the nuclear surface. The Golgi apparatus in phytoflagellates is not known to be intimately linked to the rhizoplast, where the latter has been found, but it always occurs in its near vicinity. Since striated kinetosome-linked fibrils are also common in metazoan and metaphytan flagellated cells and, as we shall see, in ciliates, it is safe to say that genesis of fibrils with a periodic structure is a common capacity of the kinetosomal and in some cases centriolar region. Their significance is in all instances unknown, but certain obvious possibilities come to mind. Considering their positions in various cells and also their suggestive resemblance to the periodic structure of collagen fibrils, one might suppose that they are elastic, supporting, or anchoring structures. Their frequent association with the nucleus suggests in addition a role in the establishment or maintenance of essential spatial and temporal relationships preceding mitosis.

The second common type of kinetosome-linked fibril is a cylindrical structure with a low-density core and dense periphery — described for the sake of convenience as tubular. Its diameter varies from 18 to 25 mμ; it is similar in size and appearance to the individual subfibrils of the flagellum but apparently is never a direct continuation of these. Lateral linkage to kinetosomes has been demonstrated clearly in *Euglena* and in *Bodo*; in several other flagellate species tubular fibrils are conspicuously present under the cell membrane, and we do not know whether they are connected to kinetosomes. Similar fibrils that appear to lack a direct kinetosome linkage compose the contractile axostyle of *Pyrsonympha* and the rodorgan of *Peranema*. Fibrils that occur in the axostyles of *Tritrichomonas, Joenia,* and *Foaina* seem to be slenderer; most authors have not given measurements, but Grassé states that they are finer, and estimates from published micrographs average about 15 mμ. All reports that include relevant information indicate that these axostylar filaments do not make contact with the kinetosome/centriole directly, but may be linked to it by striated ribbons. Since tubular fibrils occur widely among the ciliates, it will be more appropriate to reserve speculation on their significance until after examining this group.

Grimstone's (1961) tantalizing announcement of the discovery of a typical fibrillar centriole embedded in the dense hemispherical complex atop the rostral tube of *Trichonympha* is of outstanding interest. Thanks to Cleveland's comprehensive light-microscope studies (1957, 1960, and many previous papers cited therein) it would seem that metamonad zooflagellates provide a choice material for the electron-microscope study of centriolar morphology and behavior, and it had been disappointing to find no structure in the presumed centrosome regions of *Trichonympha*, *Foaina,* and *Pyrsonympha* that could be homologized with the centrioles of other cells. In all of these instances, the centrosome area is electron-dense; Grimstone's observation raises the possibility that fibrillar elements may be detected in them also if improved staining techniques are applied.

A similar dense zone was seen by Rouiller and Fauré-Fremiet (1958a) near the kinetosomes of the chrysomonad *Chromulina psammobia*; this is the only instance among the phytoflagellates for which a possible centriole distinct from a kinetosome has been reported from electron-microscope studies (although others are described in the light-microscope literature), and the question needs further study. In general — and speaking from a void, since no high-resolution studies of mitotic phenomena in flagellates have been published — it seems that, as long as the number of flagella remains small, the cluster of kinetosomes may serve all morphogenetic, kinetic, and mitotic functions required of them. As the flagellar complement increases, and as the number of associated organelles multiplies, the problems involved in establishing cell polarity and controlling the processes of division become more complex, and a division of labor in the organizational centers may be necessary. In the higher zooflagellates, a pair of " master " kinetosomes is segregated, which retain all the multi-potent generative and organizational capacities, while somatic kinetosomes are limited to the production and control of elements in their immediate environment. An electron-microscope tour through the zooflagellates, guided by the facts established in Cleveland's works, should permit us to trace the anatomical and functional segregation of kinetosomes and centrioles in this series.

In the opalinids and ciliates, polymerization of flagellar units poses the same organizational problems, with additional special features, and here, so far as we know, the answer has been rather different, since nuclei seem to have given up any morphological association with kinetosomes at any stage, and no centrioles are known.

Superorder Opalinica

The opalinids are protozoa symbiotic in the guts of cold-blooded vertebrates; they have a uniform covering of body cilia, no mouth, many equivalent nuclei, and no centrioles. The long accepted practice of classifying them as protociliates has recently fallen into disfavor. Grassé (1952), following the lead of several other workers, places them in the Superorder Opalinica of the Class Zooflagellatea. Corliss (1955) in a more recent discussion of the puzzle, favors a recognition of flagellate affinities but objects to the implication that their position is equivalent to the superorders Protomonadica and Metamonadica. In the most recent critical light-microscope study of opalinid life cycles and morphogenesis, Wessenberg (1961) finds that they share about equal numbers of characteristics with the flagellates and with the ciliates, in a combination that is unique to them. He feels that it is unjustifiable to class them with either group.

Opalinids have been examined with the electron microscope by Bretschneider (1950), Pitelka (1956), and Blanckart (1957) and finally by Noirot-Timothée (1958d, 1959), whose excellent micrographs provide most of the information for the following description. Various species of *Opalina* have been included in the studies, but no specific differences in ultrastructure are apparent. *Opalina* is elongate, obliquely rounded at the anterior end, and tapers to a point posteriorly. It is strongly flattened laterally (according to Wessenberg; some other authors have considered the flattening to be dorso-ventral). Along the oblique anterior margin a row or narrow field of kinetosomes occurs, extending part way down the ventral edge. From this field, called the falx, orderly rows of kinetosomes extend posteriad, providing the lateral surfaces with a more or less uniform covering of short cilia. Each longitudinal row of kinetosomes is called a kinety. Some kineties extend to the posterior end of the cell, while others end at any level short of this.

The electron microscope reveals that the entire surface of

Opalina bears orderly, longitudinal ridges or ribs running parallel to the kineties (Text-fig. 11). From one to 24 ribs are present between adjacent kineties — fewer anteriorly where kineties are more numerous and close-set, and more posteriorly. The number

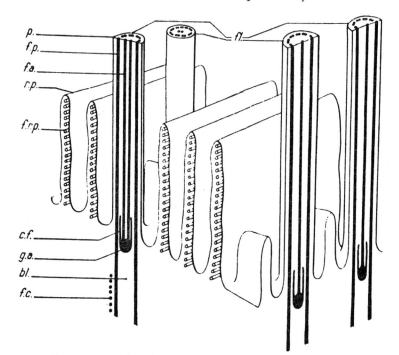

TEXT-FIGURE 11. Diagrammatic reconstruction of the pellicle of *Opalina*. fl, flagella; p, plasma membrane; f.p., peripheral fibrils of flagellum; f.a., central fibrils of flagellum; r.p., longitudinal ribs formed by folds of cell membrane over f.r.p., fibrils of the pellicular ribs; c.f., cylindrical extension from g.a., axial granule at top of bl., kinetosome; f.c., fibrils composing band comparable in position to kinetodesma of ciliates. From Noirot-Timothée, 1959.

of ribs may be reduced by fusion near the posterior tip of the cell. Ribs are about 50mμ in thickness, about 600mμ high, and separated from adjacent ribs by grooves about 100mμ wide. Each rib consists of a fold of the cell membrane enclosing a single row of 20 to 25 parallel, longitudinal, tubular fibers about 17mμ in diameter, embedded in a structureless matrix.

Grooves in which the kineties lie are wider and shallower than the others. Each cilium arises at the base of a cylindrical depression in the bottom of the groove (Fig. 67, Pl. XVIII). The two central fibrils of the ciliary axis begin in a dense spherical granule from which a short sleeve extends distally for about 100 mμ. The kinetosomes in some of Noirot-Timothée's micrographs show clearly the skewed triplet fibrils in their proximal portions.

From the anterolateral margin of each kinetosome, near its base, arises a pair of slender bands which converge and pass anteriad towards, but rarely reach, one side of the next kinetosome (whether this is the right or the left side could not be determined, although the anterior direction is clear [Pitelka, 1956]). Each band appears to be composed of short cross-fibrils. This structure, as seen in the micrographs of Pitelka and of Noirot-Timothée, thus seems to differ from the solid cross-striated and the tubular fibrils of many flagellates and ciliates, but the position of the bands in Opalina is very similar to that of kinetodesmal fibrils in some ciliates, which will be discussed in the next chapter.

The cortical cytoplasm of Opalina shows a number of remarkable differentiations of unknown significance. Running between the rows of kinetosomes, and perpendicular to them or slightly oblique, are regularly spaced zones packed with minute, rod-shaped, club-shaped, and ovoid, membrane-limited bodies. The number and sequence of these transverse zones clearly do not correspond to the number and spacing of kinetosomes. In the clear cytoplasm separating the bands are rows of larger, dense-walled, spherical profiles. Inspection of appropriate tangential sections shows that these are diverticula of the cell membrane at the bottoms of the grooves between pellicular ribs. Noirot-Timothée concluded that the diverticula were being pinched off as vesicles, and suggested a process of pinocytosis occurring here in an exceptionally orderly fashion.

Just a little farther below the surface, the bands of minute membranous bodies lose their regular orientation and spread out as a subcortical layer of varying thickness, mingled with spherical vesicles perhaps formed by pinocytosis. Below this layer the ectoplasm is extremely alveolar, containing thin-walled vacuoles of all sizes and shapes, usually with diffusely granular contents. Also in this region of the ectoplasm are abundant, very distinct

PLATE XVIII

FIG. 67. Oblique section of *Opalina ranarum.* Plane of section passes through bases of longitudinal pellicular folds and into cortex of cell. Row of cross-sectioned flagella at right. Row of kinetosomes at center. Arrows indicate fibrillar strands attached to bases of kinetosomes. × 37,500. From Noirot-Timothée, 1959.

FIG. 68. Section through endoplasm of *Opalina ranarum,* showing mitochondria (right and lower left) and cloud of fine vesicles. × 25,000. From Noirot-Timothée, 1959.

PLATE XVIII

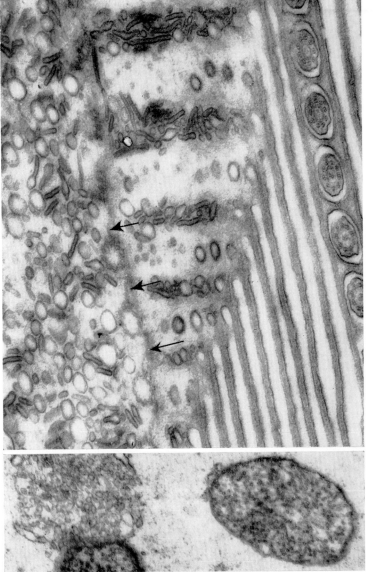

67

68

dictyosomes arranged, according to Noirot-Timothée, in two concentric strata but without any detectable connection with the kinetosomes. In the endoplasm are found the numerous nuclei, with dense peripheral nucleoli. Bodies with the characteristic structure of protozoan mitochondria but with much finer microtubules are common (Fig. 68, Pl. XVIII); Noirot-Timothée suggests that these mitochondria may be the granules described as Zeller bodies by light microscopists. In addition, clouds of tiny membranous vesicles are encountered, as well as occasional clusters or cisternae of granular membranes.

None of the electron-microscope studies has revealed any intracellular fibers of a type frequently described by light microscopists (see Wessenberg, 1961, and Grassé, 1952). Nor has the falx been identified with any certainty. In view of the large number of sections examined by the various investigators, it seems probable that falcular kinetosomes are not associated with fibers or other structures peculiar to them.

The pellicular structure of *Opalina* is reminiscent of that of the gregarines, which likewise are mouthless animal symbiotes. If absorption of nutrients can occur all over the cell membrane, the convolutions of the surface clearly would increase its effective area. However, the fibrous construction of the ribs does not make this possibility particularly inviting. The cell is elastic and flexible under mechanical distortion, but normally maintains a constant characteristic shape. Conceivably the fibrous ribs, lying perpendicular to the cell surface, might provide a degree of mechanical support while not interfering with intake of nutrients through the cell surface between them.

Electron microscopy does not yet add any fuel to arguments concerning opalinid affinities. This is not for lack of precision in information on the ultrastructure of adult *Opalina*; thanks to Noirot-Timothée's work this is relatively satisfying. Data at the electron-microscope level on nuclear phenomena, gametes, and young stages will be helpful, but only when we have similar data on many series of ciliates and flagellates to compare them with. For the time being Wessenberg's conclusion that their position is unique appears to be the most satisfying one.

K

CHAPTER 6

CILIATES

UNLIKE any other large systematic grouping within the protozoa, the Subphylum Ciliophora needs no apology. Exclusion of the opalinids from the subphylum leaves a single, almost certainly monophyletic class of ciliated organisms with differentiated macro- and micro-nuclei and no detectable centrioles. Problems of phylogeny within the group are far from being solved, but there are sufficient clues to make speculation on the subject an absorbing and rewarding game, and one to which electron microscopy should ultimately contribute in large measure. Already, ciliates have attracted a large number of investigations into their ultra-structure — because there is so much of it. Even so, the results constitute a random scattering of data which at present writing still defy systematic analysis. The very great degree of morpho-logical specialization, particularly at the surfaces, of ciliate cells means that a thorough study of a single species requires a tremen-dous investment of time. That such investments on an increasingly wider scale will pay off is not to be questioned; the structural specializations of ciliates are answers to common problems of cell function and development.

Since Volume II of Grassé's *Traité de Zoologie*, slated to include a coverage of the ciliates by Fauré-Fremiet, has not reached publication, the taxonomic arrangement to be followed here will be that of Corliss (1956, 1959a, 1961) whose work has been strongly influenced by Fauré-Fremiet (see for example the latter's brief outline of ciliate systematics in 1950). Terminology for ciliate body parts and organelles will be used as defined by Corliss (1959a).

The most striking morphological feature of ciliates is the infra-ciliature, consisting primarily of kinetosomes — whether or not they bear cilia — in characteristic topographic array. In primitive and many embryonic forms kinetosomes are aligned in meridional rows (kineties) that often become secondarily, in ontogeny or phylogeny, modified by torsion, asymmetric growth, increase or reduction in number, or incorporation of parts or entire rows into a more or less elaborate buccal apparatus. Representatives of the large Subclass Holotrichia bear (in at least some stages of their life cycle) a fairly uniform covering of body, or somatic, cilia; the buccal apparatus varies from extremely simple to rather elaborate. In some adult holotrichs and in most orders of the smaller Subclass Spirotrichia, the somatic ciliature is reduced and/or modified into compound ciliary organelles; the buccal apparatus in the spirotrichs always includes a spiraled adoral zone of ciliary membranelles.

In addition to the kinetosomes, a conspicuous feature of most ciliates is a system, or systems, of cortical or subcortical linear patterns that may in part be visible in living cells but are most clearly revealed by staining techniques. These so-called fibrillar systems are in general constant and species-specific. We shall see that a number of quite different structural entities is included, but it is necessary by way of introduction to review in the most general terms the picture presented by light microscopy (for fuller details see Gelei, 1936; Taylor, 1941; Klein, 1942; Wichterman, 1953; Corliss, 1956; and Párducz, 1958a). Silver impregnation techniques reveal the rows of kinetosomes and, typically, meridional lines connecting them like beads on a string (Fig. 69, Pl. XIX). Secondary silver meridians lacking kinetosomes also may be present. Additional parts of these silverline systems may cross-connect meridional lines or form networks between them. Other fibers arising from kinetosomes and passing deeper into the cell have been described, and a central body called the motorium or neuromotor center is sometimes reported. Argentophilic patterns associated with the buccal apparatus can become very complex.

A particularly intriguing fibrous structure, usually not argentophilic, is the kinetodesmos*, first described by Chatton and colleagues in France (Lwoff, 1950). It is a fiber accompanying

---

* This word, adapted to English via the French neologism *cinétodesme,* has been variously employed in the singular and plural as kinetodesma-desmas; kinetodesma-desma; kinetodesma-desmata. In 1959 Ehret and Powers, on sound etymological grounds, adopted kinetodesmos-desma, which usage will be followed here.

each kinety in most, but probably not all, ciliates; it typically lies to the ciliate's right (clockwise to an observer looking on the anterior pole from the outside of the cell) of the kinetosomes. This observed asymmetric relationship, stated by Chatton as the "rule of desmodexy" (Lwoff, 1950), confers asymmetry and polarity upon the kinety and upon the cell.

Many functions have been ascribed to these various structures, including those implied by the terms "neuromotor" and "neuroformative". We shall see that the evidence does not yet justify any general conclusions in this regard; some possibilities, still lacking experimental testing, are mentioned later.

None of the ciliates considered to be most primitive has been studied in detail with the electron microscope. We therefore shall select as ciliate types some representatives of the moderately advanced holotrich Order Hymenostomatida. As the most familiar of all ciliates and the one that has received most attention from electron microscopists, *Paramecium* might appropriately introduce this section. Its cortical morphology, however, is somewhat easier to understand if we consider first a group of related hymenostomes that are almost as familiar, *Tetrahymena, Colpidium,* and *Glaucoma*. The tetrahymenids not only are typical of the order but are considered by Fauré-Fremiet (1950) and Corliss (1956) to represent a possible stem group important in the ancestry of higher ciliates.

Subclass Holotrichia

Order Hymenostomatida

Tetrahymena, Colpidium, and *Glaucoma,* and species within the genera, differ in the number and arrangement of kineties and silverlines, readily demonstrated by light microscopy. At the electron-microscope level, however, no important differences are observed; the ultrastructural basis of cortical patterns is the same in all three. Early electron-microscope studies were those by Wohlfarth-Bottermann and Pfefferkorn (1953) on dried, exploded cells of *Colpidium,* and by Metz and Westfall (1954) on pellicle fragments of *Tetrahymena pyriformis,* obtained by ultrasonic bombardment of lightly fixed cells. More recently, Elliott and Tremor (1958) have reported briefly on the contact zone in conjugating pairs of *T. pyriformis,* Pitelka (1961a) has analyzed in detail the cortical structures characteristic of the group, and Roth and Minick (1961) have observed some morphological features of dividing cells of *Tetrahymena.*

The silverline system typical of all three genera is illustrated in Fig. 69, Pl. XIX. The primary meridians, except those behind the mouth, extend from pole to pole and bear the kinetosomes and smaller argentophilic granules which are "protrichocysts" or mucigenic bodies. Secondary meridians bearing mucigenic bodies but no kinetosomes typically alternate with the primaries but may be interrupted or absent; they merge with the primaries anteriorly and posteriorly. Electron-micrographs of pellicle fragments show the meridians as slender, sinuous lines with occasional expansions, often circular in outline, that represent the points of attachment of mucigenic bodies.

Examination of sections reveals that the entire surface of the cell is covered by a continuous unit membrane (Text-fig. 12). Immediately beneath this is a mosaic of flat, membrane-bounded alveoli* whose contiguous rims are the sites of silver deposition and thus define the silverline system. Each alveolus is elongate, occupying the area between a primary and the adjacent secondary meridian; transverse interruptions occur at apparently random intervals. Cilia and mucigenic bodies emerge from the underlying ectoplasm between adjacent alveoli. Ordinarily the alveoli are so flat that the space within them is almost non-existent, but under some conditions — presumably of suboptimal fixation — they become inflated and their outlines are clearly identifiable. Thus over most of the body surface the pellicle consists of three parallel

---

* In a previous description of this system (Pitelka, 1961a), the pellicular spaces were called *blisters.* Several of my colleagues have protested that this term carries pathologic connotations and hence is unsuitable for normal structures. De Puytorac (1961b), describing the equivalent system in an astome ciliate, uses the term *vesicles,* but it is employed so widely for miscellaneous intracytoplasmic spaces that confusion seems likely. *Peribasal space,* introduced by Ehret and Powers (1959) for *Paramecium,* seems unnecessarily cumbersome and descriptively inaccurate when applied to other ciliates. I am indebted to Dr. B. Párducz for pointing out to me that the structures seen are in all probability identical to those described by Bütschli (1887–1889) as superficial *alveoli.* It is a pleasure to acknowledge the priority as well as the suitability of this term.

PLATE XIX

FIG. 69. Photomicrograph of silver-impregnated *Colpidium campylum*. By B. Párducz.

FIG. 70. Section across primary meridian of *Colpidium campylum*, showing cilium emerging between adjacent alveoli (arrows) of silverline system. × 43,500.

FIG. 71. Section across secondary meridian of *Colpidium campylum*, showing continuous outer membrane and closely juxtaposed alveoli of silverline system. × 60,000.

FIG. 72. Part of pellicle fragment of sonicated *Tetrahymena pyriformis*. Empty pellicle is folded back on itself, exposing kinetosome and single kinetodesmal fibril attached to it. × 31,000. From Metz and Westfall, 1954.

FIG. 73. Pellicle fragment of sonicated *Colpidium campylum*, showing primary and secondary meridians (numbered 1 and 2, respectively) and longitudinal fibril band (3). × 11,000. From Pitelka, 1961a.

PLATE XIX

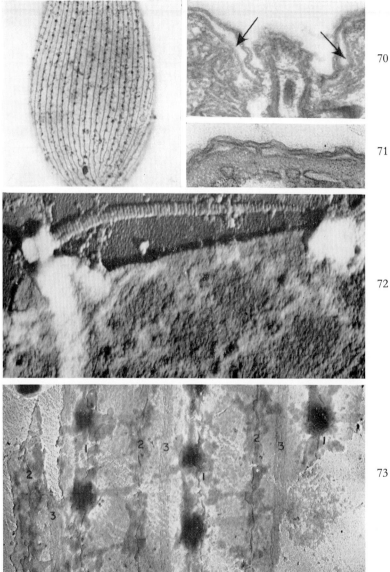

membranes, the outer one being continuous and providing the limiting membranes of the cilia, and the middle and inner layers joining to form the discontinuous but closely fitted alveoli (Figs. 70 and 71, Pl. XIX).

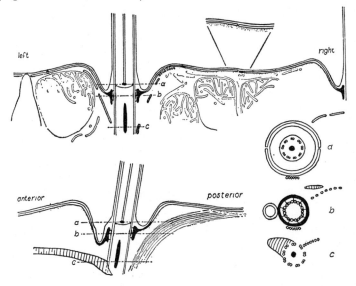

TEXT-FIGURE 12. Schematic drawings of sections through a cilium and surrounding structures in *Colpidium*. Top: transverse section showing continuous outer membrane and alveoli of silverline system. Mucigenic body in secondary meridian at left; secondary meridian at right enlarged above. To right of kineto-some are three sections of kinetodesmal fibrils and row of fibrils composing longitudinal band. To left of cilium are oblique sections of two fibrils of transverse band. Lower left: longitudinal section in an idealized plane just barely to right of primary meridian. Postciliary fibrils posterior to kinetosome; kinetodesmal fibril anterior to kinetosome. Lower right: cross-sections through cilium base at successively deeper levels as indicated by lines, a, b, and c on other diagrams. Transverse fibrils to left of (below) cilia in sections a and b. Parasomal sac and fused pellicular ring in section b. Kinetodesmal and postciliary fibril attachments to kinetosome shown in section c. From Pitelka, 1961a.

This new picture represents a considerable departure from some, at least, of the classical views of the silverline system inas-much as it is seen to be an arrangement of membranes and not of

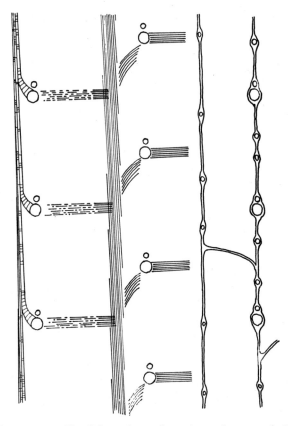

TEXT-FIGURE 13. Schematic surface view of parts of three kineties of *Colpidium,* shown at three different levels. Cilia are indicated as empty circles, with parasomal sacs shown as smaller circles anterior to them. Kinety at left is shown with kinetodesmal fibrils attached. Kinety at center is shown with accompanying longitudinal, transverse, and postciliary fibrils. Kinety at right shows outlines of pellicular alveoli along primary and adjacent secondary meridian; scattered small circles are mucigenic body attachments. From Pitelka, 1961a.

fibers. The tetrahymenids, however, do not lack fibrillar structures (Text-figs. 12 and 13). The most prominent ones are the kinetodesma (Fig. 72, Pl. XIX), shown by Metz and Westfall (1954) to consist of short, tapering, striated fibrils arising individually from kinetosomes and extending to the right and anteriad to form a loose bundle obeying the rule of desmodexy. Each kinetodesmal fibril is broadly attached to the right anterior margin of the base of its kinetosome, and extends past (but does not make contact with) one or two kinetosomes anteriad before terminating. In addition to these solid, striated kinetodesmal fibrils, three sets of fine fibrils, all tubular in appearance and about 20 mμ in diameter, occur consistently. One set is not directly associated with the kinetosomes. It lies below the innermost pellicular membrane to the right of each kinety, just above the kinetodesmal bundle. Like the latter, it is composed of overlapping short units; these collectively form a continuous band, 8 to 14 fibrils wide, that is apparent in isolated pellicle strips (Fig. 73, Pl. XIX) and evidently extends beyond the kineties to the extreme anterior pole of the cell. A second set is a transverse band of about six fibrils that arise at the left margin of each kinetosome (Fig. 74, Pl. XX) and passes up to the surface and laterally under the pellicle to terminate in the region of the next longitudinal fibril band to the left. The third set, the postciliary fibrils, arises as an oblique row of seven or eight fibrils extending from the right posterior margin of each kinetosome (Fig. 74, Pl. XX). They spread out somewhat as they pass up toward the pellicle and run posteriad to end in the vicinity of the adjacent longitudinal fibril band to the right.

Thus in these tetrahymenids each kinetosome of the somatic infraciliature is the point of origin of three diverging fibrous structures: the striated kinetodesmal fibril from its right anterior margin, the transverse fibril band from its left lateral margin and the postciliary fibrils from its right posterior margin. None of these fibrils fuses, so far as can be detected, with fibrils from any other kinetosome; *there is no fibrous connection between kinetosomes,* except in the buccal region where tracts of fine filaments do interconnect them.

Each somatic cilium of the tetrahymenid arises from a boat-shaped depression of the cell surface. The external pellicular membrane lines this depression and extends to cover the cilia. The rims of the adjacent pellicular alveoli, indented to permit egress of the ciliary fibrils, become thickened and fuse to form a continuous ring or short sleeve about the upper end of the kinetosome. Just anterior to the cilium, the cell surface invaginates in a

PLATE XX

FIG. 74. Superficial section of *Colpidium campylum*, showing one cilium, cut near its base, and two kinetosomes. M, primary meridian between alveoli anterior to cilium; T, circular profiles of transverse fibril band; P, circular profiles of postciliary fibrils; K, segment of kinetodesmal fibril from next posterior kinetosome. × 67,000. From Pitelka, 1961a.

FIG. 75. Tangential section at surface of *Paramecium aurelia*, intersecting polygonal ridges near their tops at right, more deeply at left. Each pair of cilia occupies boat-shaped space outlined by two outer pellicle membranes. A, cavities of pair of pellicular alveoli surrounding a pair of cilia; M, longitudinal silverline meridian, running horizontally in this picture; T, trichocyst tip embedded in polygon ridge; K, kinetodesmal fibril. × 22,000. From R. V. Dippell and K. R. Porter, unpublished.

PLATE XX

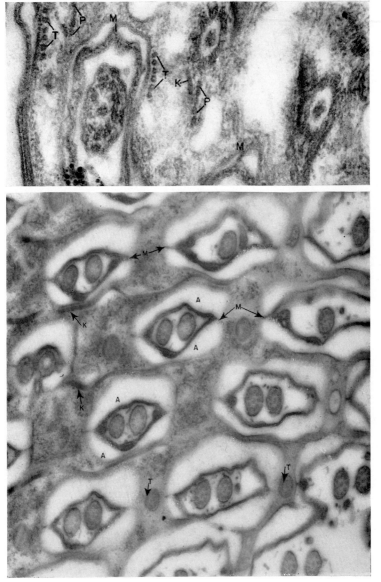

74

75

tiny finger-like diverticulum, the parasomal sac. In all three tetrahymenids studied, the central fibrils of the cilium originate in a small dense granule above the kinetosome. Within the latter, an axial rod of dense material appears, ending just short of the kinetosome's apex.

Metz and Westfall (1954) were able to isolate the buccal apparatus from sonicated *Tetrahymena* and showed that it consisted of the closely aligned kinetosomes of the three membranelles and undulating membrane, all connected to a fan-shaped array of substantial fibers apparently outlining the buccal cavity but extending beyond the mouth deep into the cytoplasm. Observations on the buccal apparatus in sectioned *Tetrahymena* are not available. In *Colpidium* and *Glaucoma,* one wall of the buccal cavity bears conspicuous ridges, and filaments that do not appear tubular in the material studied interconnect buccal kinetosomes and form a mat along one side of the cavity.

In this connection it may be mentioned that there is some question whether the fibers of the buccal cavity of *Tetrahymena* are trichites comparable to those in the pharyngeal region of gymnostomes (see below). Present evidence suggests that they are not. However in *Frontonia,* another hymenostome, Rouiller, Fauré-Fremiet, and Gauchery (1956d) have found true trichites composed of subfibrils packed in orderly arrays.

The observations of Elliott and Tremor (1958) on conjugant pairs of *Tetrahymena pyriformis* show an extremely close apposition of pellicles in the contact zone. Apparently the outer pellicular membrane, underlain by a dense cytoplasmic layer, is the only one present here, the alveoli having disappeared. The apposed pellicles are interrupted at frequent intervals by pores through which cytoplasmic continuity between mates is established. No cilia or kinetosomes are observed in the area of contact.

Relatively little attention has been given to the internal fine structure of tetrahymenids. In their comparison of dividing micronucleate and amicronucleate strains of *T. pyriformis,* Roth and Minick (1961) found a decrease in the compactness of intra-mitochondrial tubules, frequently accompanied by the appearance of an amorphous central inclusion, during early division stages of amicronucleate cells. At the same time, granular cytoplasmic membranes increased in quantity and sometimes formed compact, convoluted masses. These authors' observations on nuclear structure have been recorded in Chapter 2.

In Pitelka's studies of the three tetrahymenid genera (1961a), a layer of homogeneous dense material, varying in thickness, was commonly observed immediately beneath the innermost pellicle membrane. Endoplasmic reticulum frequently parallels this layer, and may appear as rather extensive lamellae elsewhere, often surrounding mitochondria. Some of these membranes have granular outer surfaces, and Palade particles may be very abundant in the cytoplasmic matrix. Mitochondria are distributed throughout the cell but are particularly common peripherally, where they may form compact rows between meridians. Mucigenic bodies are diffuse, low-density, pear-shaped structures with their narrow ends pressed against the outer pellicular membrane along the meridians; occasionally the outer membrane appears to be ruptured, resulting in an open pore. Curved piles of membranous discs identifiable as Golgi bodies are seen only occasionally.

In the course of their study of mucigenic bodies of the tetrahymenid, *Ophryoglena mucifera,* Grassé and Mugard (1961) observed pellicular alveoli presumably identical with those described here, although they recognized only two of the three membranes in the pellicle. A longitudinal fibril band was also seen.

The most detailed accounts of the ultrastructure of *Paramecium,* based on sectioned material of high quality, have been published by Sedar and Porter (1955), Schneider (1959, 1960a, 1960b), and Ehret and Powers (1959 and earlier papers cited there). An early study of fragmented cells by Metz, Pitelka, and Westfall (1953) provided basic information on pellicular and fibrillar patterns, while various other aspects of *Paramecium* ultrastructure have been considered by Bretschneider (1950), Tsujita, Watanabe, and Tsuda (1954, 1957), Potts and Tomlin (1955), Wohlfarth-Bottermann (1956, 1958a, 1958b), Blanckart (1957), Dippell (1958), Hamilton and Gettner (1958), and Roth (1958a, 1958b). Valuable analytical reviews of light- and electron-microscope data are provided by Ehret and Powers (1959) and by Párducz (1958a)

The polygonal sculpturing of the surface of *Paramecium* is familiar to every protozoologist; this pattern appears clearly in silver-stained specimens, which also show argentophilic lines along the kineties. The ultrastructural basis of this silverline pattern was first explained by Ehret and Powers (1959) (Text-fig. 14). The present interpretation, differing only very slightly from theirs, arises from a comparison of the published data on *Paramecium* with the later tetrahymenid studies. Cilia in *Paramecium* arise from a boat-shaped circumciliary depression in each

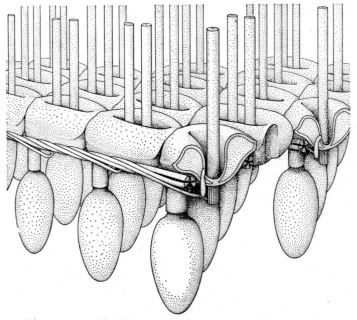

TEXT-FIGURE 14. Diagrammatic reconstruction of the cortex of *Paramecium*. Each pair of cilia emerges from the center of a polygon; the pairs of inflated alveoli defining the polygons are shown in section at the right edge of the stereogram. Parasomal sacs are shown adjacent to the cilia in these polygons. Resting trichocysts alternate with the polygons in longitudinal rows. Kinetodesmal fibers form loose cables paralleling each kinety. From Corliss 1961; redrawn from Ehret and Powers 1959.

polygon. Trichocysts emerge from the horizontal junctions between polygons in a longitudinal row. As in the tetrahymenids, the *Paramecium* pellicle consists of a continuous outer membrane overlying a mosaic of alveoli, but the latter, according to all accounts, are normally somewhat inflated. Furthermore, the alveoli are neatly dissected into units corresponding with ciliary

PLATE XXI

FIG. 76. Part of pellicle fragment of sonicated *Paramecium aurelia,* viewed from inside, showing bundles of kinetodesmal fibrils. Cilia seen clearly beyond edge of fragment at upper right, are vaguely discernible through pellicle. ×21,000. From Metz, Pitelka and Westfall, 1953.

FIG. 77. Cross-section of part of pharyngeal basket of *Nassula aurea,* showing fibrillar constitution of trichites. ×17,000. From Rouiller, Fauré-Fremiet and Gauchery, 1956d.

FIG. 78. Oblique section of elastic part of stalk of *Zoothamnium alternans,* showing striated cylinders that are ciliary outgrowths, embedded in matrix. ×14,000. From Rouiller, Fauré-Fremiet and Gauchery, 1956a.

FIG. 79. Section through part of infundibular fiber of *Campanella umbellaria.* Cross-sectioned kinetosomes in row along top, two of them linked to filaments of infundibular fiber. ×42,000. From Rouiller and Fauré-Fremiet, 1957b.

PLATE XXI

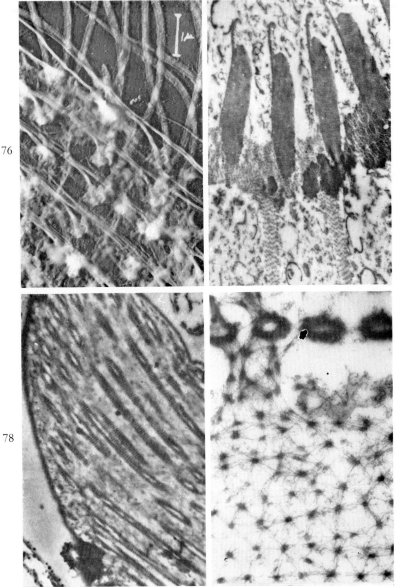

76

77

78

79

units. Each cilium (in some species or body regions, each pair of cilia) thus is surrounded by a pair of kidney-shaped alveoli defining one polygon (Fig. 75, Pl. XX). The closely apposed membranes of the paired alveoli anterior and posterior to the cilium form a double septum running between cilium and trichocysts; these are the meridional silverlines. The polygonal packing of the pairs of alveoli accounts for the polygonal silverlines, as silver is deposited along their contiguous margins or the slight gaps between them. However, marked differences in staining properties of the polygonal and meridional silverlines constitute an unsolved puzzle (Párducz, 1958a).

The ectoplasm underlying the pellicle is ridged in a pattern reflecting the polygonal packing of the alveoli. Of the fibrous structures seen in the tetrahymenids, only the kinetodesma have clear counterparts in *Paramecium*. Metz, Pitelka, and Westfall in 1953 first demonstrated that these are striated (at about 40 mμ) tapering fibrils proceeding from the kinetosome anteriad and to the right and extending along the kinetodesmal bundle for about five polygons (Fig. 76, Pl. XXI). The bundle itself may assume a loose clockwise spiral. Where cilia are paired, the single kinetodesmal fibril arises from the right anterior margin of the posterior kinetosome of the pair. The parasomal sac, recognized and named by Ehret and Powers, appears always to the right of the kinetosome — again of the posterior one if they are paired — opening to the exterior just to the right of the base of the cilium.

No fibrils corresponding in position to the longitudinal, transverse or postciliary fibrils of the tetrahymenids have been seen. However, an intriguing network of fine filaments has been demonstrated by Sedar and Porter, Schneider, and Roth. They occur parallel to but below the pellicle, at about the level of the lower ends of the kinetosomes, where they form a roughly polygonal lattice. Tracts of these filaments also appear to make contact with the innermost pellicle membrane. Their filamentous nature is too distinct in many published micrographs to support the suggestion of Ehret and Powers that they are merely tangential sections of the bottoms of pellicle membranes. Roth (1958a) reported that these, as well as a mat of similar filaments under the membrane of the buccal cavity, were tubular fibrils with a diameter of about 20 mμ; Schneider (1959) described them as threadlike filaments with a diameter of 5 mμ. The polygonal meshwork was called the infraciliary lattice system by Sedar and Porter, after a pattern discovered in the same position by the light microscopist, G. von Gelei (see Párducz, 1958a).

Ehret and Powers (1957, 1959) have devoted considerable attention to the fine structure of the buccal cavity of *Paramecium bursaria*. They were able to show, in confirmation of earlier light-microscope work, that the buccal apparatus includes three long columns of cilia, with four rows of regularly close-packed cilia in each column, and a non-ciliated ribbed wall. In the ciliated columns, one parasomal sac is present for each four cilia, and fibers identified as kinetodesmal (but looking rather filamentous in the published print) run beneath rather than beside the crowded kinetosomes. The ribbed wall is heavily ridged and underlain by fibrous structures of uncertain arrangement.

The morphology of discharged trichocysts of *Paramecium* was described in Chapter 2. The undischarged trichocyst consists of an elliptical or carrot-shaped body of rather low density, surrounded by a membrane. This is surmounted by a bullet-shaped, heavy-walled cap that is pressed against the outer pellicle membrane between alveoli. Within the cap is a slender dense rod that becomes the pointed tip of the discharged trichocyst. In dividing cells, fibrous elements are distinguishable in the dense walls of trichocyst caps cut in cross section. The fibers, according to Ehret and Powers, resemble the axial fibrils of the cilium, but their outlines seemed considerably less clear and accurate counts of their number were not possible. Trichocysts in at least some ciliates are derived from kinetosomes, according to Lwoff (1950). Additional observation on their development and fine structure is needed.

The cytoplasmic matrix in *Paramecium* (Fig. 75, Pl. XX) is rather densely packed with Palade granules, meandering double membranes or vesicles, both smooth and granular, and mitochondria of conventional microtubular structure — but no Golgi bodies have been reported. In addition several peculiar details have been described. Schneider (1959, 1960a, 1960b) illustrates tubular structures, sometimes in rather loosely parallel tracts and sometimes very precisely packed; these he finds only in the region of the complex contractile vacuole, described in Chapter 2.

Elsewhere rather large (up to 7 μ in diameter) masses of convoluted, smooth-surfaced double lamellae are infrequently seen; lipid droplets often are present nearby. The significance of these membranous structures is unknown. Nuclear morphology has been described in Chapter 2.

Kappa, the DNA-containing cytoplasmic particle responsible for the killer trait in *Paramecium aurelia,* shows, according to Dippell (1958), similarities to bacteria or rickettsiae. Each kappa particle is surrounded by a double membrane and filled with a diffusely filamentous material enclosing small, dense, irregular granules. Some kappa particles contain an inclusion consisting of concentric cylindrical membranes surrounding a granular core. Beale and Jurand (1960) conclude similarly that the mu particles of another killer strain of *P. aurelia* are bacteria modified somewhat by a long history of obligate symbiosis.

In summary, most of the structures seen by light microscopists have now been identified in electron micrographs of *Paramecium* and the tetrahymenids. The pellicle comprises three membranes — the outer continuous one and the middle and inner ones outlining the alveoli of the silverline system. Three morphologically distinguishable kinds of fibrils are seen: (1) the solid, striated kinetodesmal fibrils; (2) the very orderly tubular, 20-mμ fibrils of the tetrahymenids; and (3) the filaments of the infraciliary lattice of *Paramecium* and of the buccal zone in both *Paramecium* and the tetrahymenids. According to Roth, the last two categories would be the same, but micrographs showing a tubular nature for the filaments in cross-section are not available, and their random packing and tendency to appear sinuous give them a decidedly different appearance from the very precisely arranged tubular fibrils of group 2. Conversely, it hardly needs emphasizing that fibrillar structures that look alike in electron micrographs are not necessarily alike in any other property.

Order Gymnostomatida

Embarking on a systematic tour of the ciliates we meet first a group of great significance, the Order Gymnostomatida. According to Corliss (1956, page 81), "The most primitive forms [of gymnostomes], with their apically located cytostome, simple axis of symmetry, and uniform somatic ciliature, may well be considered the prototype" of the Ciliata. Present evidence strongly implicates them in the ancestry of trichostomes, chonotrichs, and hymenostomes, at least. Thus some of the ultrastructural enigmas of higher forms might be clarified by scrutiny of the simplest sorts of gymnostomes. Unfortunately, only a few tantalizing bits of information are yet available in published form, all from the Paris laboratory of Fauré-Fremiet and Rouiller. So far, this all concerns the morphology of pharyngeal protein fibers that in the carnivorous gymnostomes appear as an expansible circlet of rods known as trichites, and in the herbivorous gymnostomes as a rigid pharyngeal basket. Rouiller, Fauré-Fremiet, and Gauchery (1956d) show that all of these fibers are firm and elastic, birefringent, and composed of elementary fibrils aligned in a paracrystalline order. The elementary fibrils have a diameter of 15 to 20 mμ, appear tubular in cross-section, and are packed in evenly-spaced parallel rows within each trichite, thus being reminiscent of the axostyle fibrils of Grassé's *Pyrsonympha*. They are not, however, contractile.

In the carnivorous *Coleps hirtus* the trichites are spaced in a discontinuous cylinder around the buccal depression. A similar picture, with some variation in the orderliness of fibril packing, is reported for two species of *Prorodon*. The pharyngeal basket of the herbivorous *Nassula aurea* contains 25 trichites, elliptical in cross-section, composed of densely and very regularly packed elementary fibrils (Fig. 77, Pl. XXI). From one side of each rod a few rows of fibrils extend medially as a curved lamella; on the outer side the rods are united by a sheath consisting of loosely arranged elementary fibrils, and from this in turn regular humps or crests of the same composition protrude into the cytoplasm. Here and there within the sheath local zones of orderly packing are observed. The authors state that a more complicated sort of basket is observed in *Chlamydodon* and in *Dysteria*.

Order Suctorida

The Order Suctorida is a compact group of ciliates characterized by the absence of cilia in the adult stage and the presence of tentacles used for the capture and ingestion of living prey. Two types of tentacles may be present: prehensile ones capable of attaching and holding large, active prey, and sucking ones through which protoplasm of food organisms, or sometimes entire organisms, flows into the suctorian's body; or both functions may be performed by the same tentacles. Asexual reproduction is by internal or external budding of ciliated larvae which swim away and shortly metamorphose into sessile, non-ciliated adults. Kinetosomes, scattered or in groups, are present in the adult.

Suctorians present at least two problems of broad general interest. One is the morphological basis of tentacle action: by what means are prey organisms held and immobilized, and how is their cytoplasm moved into the predator's body? The second is the morphogenesis of the ciliated larva, and in turn of the tentacled adult. A limited amount of electron-microscope data bearing on these problems is available. Observations on dried whole tentacles by Blanc-Brude, Dragesco, and Hermet (1951) and by Rudzinska and Porter (1954a) provided suggestive preliminary information, augmented by Rouiller, Fauré-Fremiet, and Gauchery (1956b) from detailed examination of thin sections of the two types of tentacles of *Ephelota plana*. A continuing study of *Tokophrya infusionum* by Rudzinska and Porter (1955) has led to thorough accounts of macronucleus and contractile vacuole ultrastructure, already discussed in Chapter 2, but their other data, scattered through several unillustrated notes, have yet to be assembled in a documented report. An abstract by Pottage (1959) deals with aspects of morphogenesis in *Discophrya piriformis*.

In *Ephelota*, the cytoplasm of the adult cell body is limited by a thick (to $0·4\,\mu$), smooth, homogeneous layer called the epiplasmic membrane, overlain by an alveolar cuticle consisting of packed spheroidal vesicles of various sizes. The epiplasm is shown by light microscopy to be protein, while the cuticle appears to be predominantly mucopolysaccharide. Fibrous cylinders identifiable as resting kinetosomes are seen erratically, embedded in the epiplasmic membrane, perpendicular to its surface. Over the tentacles the alveolar layer continues without interruption and the epiplasmic membrane narrows to form a thin pellicle, sometimes convoluted. The sucking tentacles are extensile and contractile; they average $3\,\mu$ in total diameter. Within the epiplasmic membrane a cortex of cytoplasm surrounds a layer of close-set, longitudinal, tubular fibrils, about 20 mμ in diameter. From this layer 26 to 28 radially disposed crests or lamellae, composed of similar longitudinal fibrils, extend into the core of the tentacle. The core in the non-feeding ciliates examined appears to be occupied by undifferentiated cytoplasm.

Prehensile tentacles of *Ephelota* are band-shaped, up to 10 μ in width, and taper to a pointed tip; they are capable of slow movements of flexion and contraction. Their pellicle and cuticle are like those of sucking tentacles, but here the resemblance ends. Internally the prehensile tentacle is subdivided by septa continuous with the pellicle into five to seven longitudinal compartments. Within each compartment an axial fiber (detectable by light microscopy and protein in nature) consists of a large bundle of very fine (no dimensions given), approximately parallel, longitudinal filaments, embedded in a somewhat granular matrix. In cross-section the disposition of fibrils within the bundle is seen to be irregular, enclosing empty spaces in a sponge-like mass. The bundles extend well into the cell body where they are surrounded by endoplasm.

The very differently arranged fibrillar structures in the two types of suctorian tentacles invite speculation, but this is better deferred until we have examined fibrous structures in additional ciliates. That suction operates in suctorian feeding is indisputable (Kitching, 1954) but its nature remains unknown. Peristaltic waves of contraction in feeding tentacles have been described, but some authors consider these insufficient to account for the flow of food. Kitching has noted a wrinkling of the cuticle over the cell body following capture of prey and suggests an expansion of the body surface with resulting negative internal pressure. If the cuticle resisted inward collapse, suction via the tentacles would result. Unfortunately, no published electron-microscope studies have included detailed observations on feeding animals.

Within the suctorian body, the food is enclosed in vacuoles, according to Rudzinska and Porter (1954b). Rudzinska (1958) found the surface of *Tokophrya* to be limited by two membranes separated by a narrow space. In some of her micrographs the outer one seems to be double, but whether this is comparable with the pellicle of tetrahymenids is not evident. A zone of homogeneous, moderately dense material appears outside of the membranes, but nothing comparable to the alveolar layer of *Ephelota* is seen. A silverline system is present in both adults and larvae but has not been studied with the electron microscope.

Rudzinska's study of the contractile vacuole region in *Tokophrya* showed that a group of cilium-less kinetosomes was always present near the pore of that organelle. An invagination of the pellicle marking the beginning of brood pouch formation appeared next to this kinetosome field. Pottage (1959) states that scattered kinetosomes are present in the ectoplasm of adults of *Discophrya piriformis*. They also are seen in the brood pouch wall, and of course in the ciliated larva. No trace of associated kineto-desma or other fibrils was observed by him. However, one of Rudzinska's (1958) micrographs of the kinetosome field in the *Tokophrya* adult shows distinct connecting fibrils. From Pottage's description, the cortex of the adult *Discophrya* seems to resemble that of *Ephelota*. Tentacles are reported to contain 25 to 30 fibrils arranged in ten groups of two or three each, forming the wall of the internal canal, which continues deep into the cell body. The larva has in its cytoplasm tubes similar to these intracytoplasmic portions of the tentacle canal, and Pottage suggests that tentacles develop during metamorphosis by outgrowth from these tubes.

Mitochondria in the suctorians are of typical microtubular construction. Elements of the cytoplasmic matrix are conventional, but no Golgi bodies have been found. Rudzinska (1958) showed peculiar stacks of very small, somewhat inflated, membranous discs that occurred commonly through all parts of the cell; if these were flatter they would resemble minute dictyosomes.

Attachment of the adult suctorian to a substrate may be by means of a stalk terminating in a firmly adhesive disc. Rudzinska and Porter (1954a) found the intact dried disc to consist of a potlike mass of branching fibrils, the finest unit apparently about 15 mμ. Larger branches appeared to be formed of bands or cables of finer units.

Order Chonotrichida

Only a single report in the electron-microscope literature deals with ciliates of the Order Chonotrichida. Like the suctorians, these ciliates are sedentary as adults, with a reduced somatic ciliature, and have ciliated migratory young. The adult attaches to the substrate frequently by means of a non-contractile stalk, which in *Chilodochona* (Fauré-Fremiet, Rouiller, and Gauchery,

1956c) is composed primarily of protein. Light microscopy reveals converging fibrils and a cluster of vesicles in the cell at the point of origin of the stalk. In electron micrographs, these vesicles are seen to be rather thick-walled, vase-shaped "glandules" embedded in the cytoplasm and opening into the cavity of the stalk. The contents of the glandules, probably shrunken during preparation, are dense and amorphous but are prolonged into the stalk as discrete, dark fibers with a diameter of 60 to 120 mμ. In the cytoplasm around the glandules are numerous, very dense, spheroidal bodies that conceivably represent precursors of the protein stalk secretion. Below a protoplasmic lip surrounding the origin of the stalk, its surface appears to be formed by a coalescence of peripheral fibers, with a dense, serrated external layer. Fibers within the stalk are rather closely packed, with interconnecting strands that may represent a precipitated matrix material.

These observations are of particular interest inasmuch as they demonstrate the existence and the morphology of a specialized secreting organelle in a protozoan cell. The glandules are totally different from the structures from which stalks in some other ciliates (e.g., peritrichs, see below) take origin. The authors cite their unpublished data, however, to the effect that certain gymnostomes believed to be the closest relatives of chonotrichs accomplish temporary attachment by means of adhesive strands secreted in simple glandules resembling those of *Chilodochona*.

Order Trichostomatida

In a speculative scheme of ciliate phylogeny, Corliss (1956) recognized the Order Trichostomatida as a polyphyletic assemblage and placed it on the highway leading from gymnostomes to hymenostomes; all other orders thus far considered occupied side roads. In more recent studies (1958, 1961) he concludes that the pivotal hymenostomes more probably arose from certain gymnostomes and that the trichostomes constitute instead one or more culs-de-sac. Be that as it may, certain trichostomes exemplify with beautiful clarity the ontogenetic progression from primitive gymnostome-like young to highly differentiated adult forms (see Fauré-Fremiet, 1950) and as such invite critical electron microscope scrutiny. Unfortunately, work on typical species has not progressed far enough to justify inclusion here.

The only trichostome for which limited electron-microscope data are available is *Isotricha,* representative of a family living symbiotically in the rumina of ungulate mammals. Bretschneider (1950) and Noirot-Timothée (1958b, 1958c) have investigated the peculiar layer that forms a distinct boundary between ectoplasm and endoplasm, visible in living and stained cells in the light microscope. Noirot-Timothée's electron micrographs show that it is composed of two layers of very fine parallel filaments (diameter not given). In all cases filaments in the two layers are oriented perpendicular to each other. From the internal layers, tracts of filaments pass inward and anastomose to form a net, like a string bag, around the nucleus. This complex, called the karyophore, appears to act as a nuclear suspensor. The boundary layer separates a rather homogeneous, finely filamentous ectoplasm from an inclusion-filled endoplasm. On both sides of the filamentous layer and along the fibers of the karyophore are seen abundant, small, disc-shaped bodies, about 350 mμ in diameter, apparently membrane-limited and filled with a variably dense granular material. The filamentous layers, Noirot-Timothée suggests, provide an elastic support for the body, which is notably plastic but not contractile.

Noirot-Timothée notes that the kinetosome structure in *Isotricha* is conventional, with a basally open fibrous cylinder capped by a cupped diaphragm above which the axial granule and the central ciliary fibrils appear. From the base of the kinetosome, a short root fiber descends directly into the cytoplasm.

Order Astomatida

The Order Astomatida presently is something of a waste-basket for symbiotic (mainly in annelids) holotrichous ciliates whose lack of any mouth structure obscures their true affinities. Trichostome, hymenostome, apostome, and thigmotrich relationships have been recognized for various astomes in recent years, and eventual abandonment of the order seems likely. Such a step is favored by a current authority on the group, de Puytorac, who has used the electron microscope to examine several species (1952–1961b).

Metaradiophrya gigas is one of a group considered by de Puytorac to be derived from thigmotrichs. It is a large, uniformly ciliated organism bearing an anterior hooked holdfast apparatus and a cortical system of longitudinal fibers assumed to be skeletal. It has a silverline pattern of rather irregular polygonal plaques, and a ridged ectoplasm. The troughs between ridges are, according to de Puytorac, occupied by membrane-limited alveoli, overlain by a continuous outer membrane. Silver is deposited, just as in the tetrahymenids, along lines separating adjacent alveoli.

Light-microscope study of *Metaradiophrya* shows that the 65 skeletal rods, which attach anteriorly to the holdfast, parallel the kineties on their right sides and that each consists of a solid, heavy fiber tapering as it extends posteriad and finally being replaced by a bundle of smaller fibers. They are anisotropic, and are composed of protein. Additional systems of fibers attached to the holdfast are assumed to be contractile. The electron microscope reveals a pattern of remarkable complexity. The posterior portions of the skeletal fibers consist of dense fibrils arising individually from kinetosomes along the kinety and passing to the right and anteriad to join a loose bundle of overlapping units, exactly like the kinetodesma of *Paramecium* and the tetrahymenids. In *Metaradiophrya* the tapering kinetodesmal fibrils are relatively long; the bundle at any one level contains seven to 15 of them. Beginning at a specific level, successive new kinetodesmal fibrils, instead of remaining separate, fuse with each other to form a homogeneous solid rod of steadily increasing diameter. For some distance, the diminishing bundle of fibrils left over from more posterior kinetosomes runs parallel to the widening solid rod. Anteriorly, some of the kinetodesma reach the shaft of the holdfast, which is composed of a similar dense material. Kinetodesmal fibrils, the fused rod, and the holdfast all show a periodic banding at about 20 mμ, but the relative width of light and dark stripes within the period differs in the smaller and larger fibers. The pellicle over the shaft and surrounding the protruding hook of the holdfast is at least double-layered, and fine longitudinal fibrils are arrayed against its inner surface.

In addition to the kinetodesmal fibrils, each kinetosome is the point of origin of other fibrous elements. In parts of the anterior region of the body, these have the form of tubular fibrils, about

15 mμ in diameter, that aggregate to form dorsal and ventral myonemes attaching to, and presumably serving to move, the holdfast. In some sections the fibrils appear to be grouped in evenly-spaced parallel arrays. Elsewhere their disposition seems less regular and their paths sinuous; this latter effect might conceivably be a fixation artifact. De Puytorac believes that each tubular fibril is composed of two or more protofibrils, about 5 mμ in diameter, wound together as a helix.

Over most of the body of the cell, a complex and orderly filamentous layer separates ectoplasm from endoplasm; the filaments originate at the bases of the kinetosomes and comprise three distinct sets. Most superficially located are slender packets of filaments that pass transversely to connect kinetosomes of adjacent kineties. Just below these, and to the right of each kinety, is a longitudinal, flattened, filamentous band. Also parallel to the kineties but alternating with them are larger filamentous cords.

De Puytorac suggests that at least the external, transversely oriented component of his system of filaments is comparable to an ectoplasm-endoplasm boundary he has seen in other astomes, and also to the filamentous boundary layer reported by Noirot-Timothée in *Isotricha,* although no connections with kinetosomes were noted by her.

In addition to the fiber systems, de Puytorac comments on the presence of abundant, tiny, rod-, disc-, or dumb-bell-shaped bodies in the ectoplasm, mitochondria concentrated in the peripheral endoplasm, and rather extensive arrays of concentric granular endoplasmic reticulum. The remarkable structure of contractile vacuole pores has been discussed in Chapter 2.

It is particularly regrettable that ciliates of the Order Apostomatida have not been examined with the electron microscope, since they exemplify with unsurpassed clarity the involved but orderly processes of ciliate morphogenesis. It was largely on the basis of the monographic studies of apostomes by Chatton and Lwoff (1935) that the influential French school of protozoologists first formulated their concepts of the importance of kinetosomes in development (see Lwoff, 1950). The curious, symbiotic thigmotrichs also have been neglected by electron microscopists.

Order Peritrichida

The spectacular contractility of peritrich ciliates has earned them repeated notice, but little as yet in the way of concerted study, from biologists interested in the chemistry and physics of protoplasmic contraction. The cell body of the peritrich is commonly goblet- or bell-shaped. The flat upper or oral surface consists of a disc usually called the peristome, bearing peripherally two or three parallel, spiral rows of cilia that lead via a more or less profound canal, the infundibulum or buccal cavity proper, to the cytostome. The aboral region bears a girdle of locomotor cilia called the trochal band in motile species and in migrant individuals; in adults of sessile species an aboral polar area called the scopula produces an adhesive substance, often forming a long stalk that encloses a myoneme in some species.

Rapid protoplasmic contraction is observable in several parts of the body: a protoplasmic collar around the peristome may constrict to enclose peristome and its cilia; the body itself may contract into a tight ball; a basal disc in creeping motile species may contract, and finally and most flamboyantly, the long stalk may contract abruptly into a tight coil. The latter is true only of some genera; others have non-contractile stalks. The myonemes observable in stalk and body by light microscopy color intensely with protein stains and are optically anisotropic, showing the same alterations on contractions as metazoan muscles (Fauré-Fremiet, Rouiller, and Gauchery, 1956b). Myosin-like proteins have not been demonstrated in them. Levine (1960) has shown specific ATP-ase activity in the stalk and body myonemes of *Vorticella*. But stalk models differ from muscle in details of their response to ATP (Hoffmann-Berling, 1958a, b).

Recovery from contraction requires an antagonistic system of some kind, and the peritrichs appear to be provided with elastic skeletal structures that serve this purpose.

Comparison of the ultrastructure of contractile and non-contractile peritrich stalks provides a basis for unassailable identification of myonemes versus supportive elements. Pertinent information comes from the work of Fauré-Fremiet, Rouiller, and Gauchery (1956b), Randall (1956, 1959a, 1959c), Rouiller, Fauré-Fremiet, and Gauchery (1956a, 1956c), and Sotelo and

Trujillo-Cenoz (1959). The non-contractile stalk in *Campanella, Opercularia,* and *Epistylis* consists of a cylindrical bundle of uniquely modified cilia, surrounded by a homogeneous sheathing membrane apparently elaborated by an ectoplasmic lip around the scopula. Sections through the scopula of *Campanella* parallel to its surface show concentric rings of typical basal bodies. Conventional cilia, complete with 11 fibrils and membrane, arise from the kinetosomes and are clearly recognizable as such in sections cut near the cell end of the stalk. Their diameter increases rapidly, so that their membranes come into contact with one another and shortly their axial fibrils disappear. What remains then is a bundle of ciliary membranes forming tubules about 350 to 400 mμ in diameter. Longitudinal sections demonstrate that the membranes have become fibrous; a diagonal lattice of criss-crossed fine filaments (5 to 10 mμ in diameter) is evident. However, apparent continuity of filaments through adjacent tube walls suggests that a matrix material may contribute to this structure.

In *Opercularia* the situation is similar, but here it appears that the peripheral fibrils, rather than the membrane, persist in the stalk, embedded in a low-density matrix. The fibrils in this case acquire a distinct transverse striation, with a period of about 22 mμ near their origin and up to 44 mμ distally. In *Epistylis* the stalk is again composed of packed tubules; here they appear as thin-walled cylinders except near the base (the earliest part formed) of the stalk where nine fibrils are apparent in their walls.

The contractile stalks of the vorticellids contain an elastic substance that may surround and enclose the myoneme or may simply parallel it. The elastic element of *Zoothamnium* is composed of striated cylinders that are ciliary in origin, embedded in a homogeneous matrix (Fig. 78, Pl. XXI). *Carchesium* stalks likewise contain banded fibers that are ciliary outgrowths. In *Vorticella* the fibrils are few in number, unbanded, and attached throughout their length to the sheathing membrane of the stalk.

The contractile myoneme in these species, when examined at high resolution, is seen to consist of a bundle of very fine filaments 2 to 4 mμ in width and of indefinite length. In cross-section the bundle appears alveolar, enclosing many circular spaces identified by Sotelo and Trujillo-Cenoz as a system of membranous canaliculi.

This stalk myoneme in vorticellids is directly continuous with a system of intracellular myonemes that presumably function in body contraction. Such myonemes occur in most peritrichs; near the oral end of the cell they lie immediately below the ectoplasm; aborally they converge to insert on the scopula in *Opercularia*, to insert on skeletal elements of the basal disc in mobile forms (see *Trichodinopsis*, below), or to continue into the stalk in vorticellids (Rouiller, Fauré-Fremiet, and Gauchery, 1956c; Fauré-Fremiet and Rouiller, 1958b; Sotelo and Trujillo-Cenoz, 1959). In all instances they are composed of very fine filaments, permeated by a system of fine branching canaliculi and small vesicles. The periphery of the myoneme bundle is never completely bounded by membranes, but often is partially sheathed by discontinuous flattened or inflated membranous sacs that on their cytoplasmic side may be studded with Palade particles. The membrane systems within and around the myonemes appear to be at least intermittently in continuity with each other and also with endoplasmic reticulum of the general cytoplasm. Both the French group and Sotelo and Trujillo-Cenoz point out the parallelism between this and the sarcoplasmic reticulum of vertebrate striated muscle (a canalicular system around and in the Z-bands of muscle fibers).

Fauré-Fremiet, Rouiller, and Gauchery (1956b) have observed myonemes, with the same structure, associated with the cytoplasmic collar at the rim of the peristome, or with the peristomal disc itself. The same authors state that restoration of body shape following contraction is assured by a resistant cuticle underlain by annular fibers, visible in appropriate thin sections as single, regularly spaced, slender, dense threads.

This survey indicates that a fairly impressive system of co-ordinated contractile structures and skeletal elements is present in most peritrichs. But what has been glimpsed so far is simple by comparison with the incredibly elaborate architecture of a mobile peritrich, *Trichodinopsis paradoxa*, investigated by Fauré-Fremiet, Rouiller, and Gauchery (1956a). This species has a basal disc supported by a complex skeletal armature, consisting of circular and radial elements, and bordered by a delicate cytoplasmic rim, the velum, and several rows of locomotor cilia. The prominent circular skeletal component is made up of heavy, dense, homogeneous plates overlapping to form an articulated ring. Overlying this is a corona of very regularly placed, slender rods radiating from the central region of the disc out past the ring to the ciliary row overlying the velum. These are dense and homogeneous like the ring; cytochemical techniques reveal that both skeletal elements are made up of scleroproteins.

Several filamentous tracts assumed to be contractile are inter-woven among the structures described (Fig. 80, Pl. XXII). These variously ramify between the kinetosomes of the locomotor cilia and the pellicle of the basal disc, between the latter and the anterior ectoplasm, and between the basal disc pellicle and the ring and radial skeleton. In addition an annular myoneme, of a structure heretofore not encountered, is closely applied against the pellicle of the basal disc just below the velum. It consists of a compact band of smooth parallel laminae, running obliquely within the circular unit, embedded in a matrix of moderate density. The Fauré-Fremiet group thus has distinguished two types of myonemes in ciliates: the finely filamentous endoplasmic myoneme and the laminated ectoplasmic myoneme. Both types will be encountered again in the heterotrichs.

The long tubular infundibulum leading to the mouth of *Trichodinopsis* is supported by three elastic, protein fibers. Each of these consists of a bundle of regular, parallel fibrils, about 30 mμ in diameter. Some micrographs suggest a close relationship with kinetosomes of the peristomal ciliature, but details are not discernible.

An infundibular skeleton of very different construction is present in *Campanella umbellaria* (Rouiller and Fauré-Fremiet, 1957b). It is a long, coiled fiber with a maximum diameter of about 2 μ. In thin sections it is seen to be composed of a three-dimensional hexagonal lattice of fine (1–2 mμ) filaments, with small granules at the nodes (Fig. 79, Pl. XXI). On the cytoplasmic periphery, the lattice is often flattened into compact parallel rows or small clusters of nodes, while on the medial side it shows definite connections with the kinetosomes of the infundibular ciliature. From the base of each kinetosome a bundle of connecting filaments extends toward the main fiber; nodes are arranged in parallel rows across this bundle.

Additional observations by Rouiller and Fauré-Fremiet (1958b) have to do with the ciliary apparatus of *Ophrydium versatile,* in

PLATE XXII

FIG. 80. Section of *Trichodinopsis paradoxa,* cut perpendicular to plane of basal disc. Very dense skeletal elements at right; bases of cilia of locomotor fringe at left. Fibers assumed to be contractile interconnect other organelles. × 10,000. From Fauré-Fremiet, Rouiller and Gauchery, 1956a.

FIG. 81. Superficial section of trochal band of *Ophrydium versatile* migrant, showing oblique rows of kinetosomes, with attached fibrils, cut at successively deeper levels from left to right. × 39,000. From Rouiller and Fauré-Fremiet, 1958b.

Plate XXII

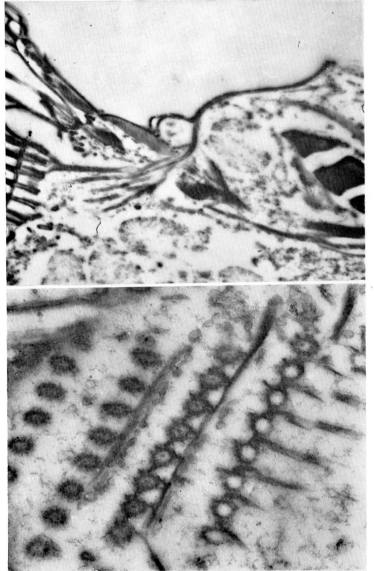

80

81

both migrants and sessile adults. In the migrant the cilia of the trochal band arise from kinetosomes aligned in short oblique rows within the band encircling the cell. The kinetosomes, of conventional structure, are directly linked to one another along each row by fibrous bridges (Fig. 81, Pl. XXII). At the same level a continuous heavy fiber or band of filaments parallels the row along one side and is attached to each kinetosome. At a deeper level a fibril arises from each kinetosome and passes in a direction perpendicular to the oblique row. The length and course of these fibrils are unknown, but the authors suggest that they may represent kinetodesma and in the original print supplied by Prof. Fauré-Fremiet traces of cross-banding are detectable. In addition to the fibrous structures, rows of tiny vesicles and tubules are aligned close to the oblique fiber paralleling each row. Adult cells have no aboral cilia and a reduced number of kinetosomes in the aboral band, and these lack associated fibers entirely. They seem to be open apically, ending at or slightly removed from the inner surface of the pellicle. Granules appear in the core of the kinetosomes in this stage.

Subclass Spirotrichia

Order Heterotrichida
   Among the six orders of the Subclass Spirotrichia, the Heterotrichida occupy a central position comparable to that accorded the gymnostomes and hymenostomes among the Holotrichia. Most heterotrichs have a holotrich-like uniform somatic ciliature, but all display the spiraled adoral zone of membranelles that is the major distinguishing feature of the subclass. The buccal cavity typically is a broadly expanded funnel, the peristome. In recent years the esthetically delightful heterotrich, *Stentor,* has achieved a status almost rivaling those of *Paramecium* and *Amoeba* as an experimental organism (see Tartar, 1961). This is largely because of its extraordinary amenity to surgical methods that in the hands particularly of Tartar and of Weisz (1954, 1956) have yielded invaluable information on morphogenesis, but also because of its phenomenal contractility and extensibility. Since *Stentor* has a precise and intricate cortical architecture whose main features are readily observed even at

low magnifications of the light microscope, it is a choice subject for combined experimental and ultrastructural studies. The first necessary approach has been made in the critical, detailed electron-microscope investigation of *S. polymorphus* by Randall and Jackson (1958).

Some species of *Stentor* are colored by pigment granules distributed in regular longitudinal rows. *S. polymorphus,* lacking pigment, has alternating granular ridges and clear furrows in which lie the kineties. When attached to a substrate by its posterior end and fully extended in the familiar elongate trumpet shape, *Stentor* may reach a length of 2 mm. When detached and swimming, its shape is conical and its length about 0·7 mm. Maximal contraction reduces it to a spheroidal body about 0·25 mm in length. Its cell volume in the superextended state may be four times that at full contraction. According to Randall and Jackson, chemicals that inhibit contraction also interfere with adequate fixation, so all electron micrographs are of sections of animals at least partly contracted by the initial impact of the fixative.

The cell surface, in addition to being longitudinally ridged, is horizontally wrinkled in fixed cells. In most areas it consists of at least two unit membranes, or possibly of a single one underlain by a layer of very small flattened alveoli (smaller and less uniform than those composing the silverline mosaic of the hymenostomes; unlike the latter, the *Stentor* pellicle disintegrates completely under sonic bombardment or detergent treatment). Near the buccal cavity, however, two outer membranes are elevated in scallops above the ridged ectoplasmic surface which is bounded by one, or possibly two, additional membranes. The picture in transverse sections here is very like *Paramecium*. Another figure reveals the presence of a parasomal sac adjacent to a cilium base. The cytoplasm of *Stentor* is rather hyaline, with scattered small particles and very abundant vesicles and vacuoles. Many of these, ranging up to 2 μ in size and bounded by unit membranes, fill much of the endoplasm. Just below the buccal cavity is a zone packed with vacuoles some of which have two parallel limiting membranes. These vacuoles increase in size some distance below the buccal cavity; Randall and Jackson suggest that they may be formed at that surface and migrate inward. Considerable intake of water occurs when *Stentor* becomes fully extended; the vacuoles suggest a mechanism for this intake.

Arising at apparently regular intervals from the inner surface of the buccal cavity membrane and passing posteriad in the scant cytoplasm between vacuoles is a system of fine tubular fibrils aligned in sheets (Fig. 82, Pl. XXIII). The fibrils are about 25 mμ in diameter and in section resemble fibrils of the membranelle roots (see below). Their terminus is unknown.

Cilia and kinetosomes are of conventional structure. Often the kinetosome appears to be closed off at the bottom where attached fibers arise. In addition, many kinetosomes — most frequently in conjugant or dividing cells — contain in their lumina very dense granules, about 25 mμ in diameter, aligned in neat rows parallel to the kinetosome axis.

Somatic kineties have microscopically visible kinetodesma. In the electron microscope (Fauré-Fremiet, Rouiller, and Gauchery, 1956b; Fauré-Fremiet and Rouiller, 1958a; Randall and Jackson, 1958; Inaba, 1959) the fibers are seen to consist of up to 24 parallel longitudinal ribbons, each ribbon composed of 24 to 30 fine tubular fibrils about 20 mμ in diameter, joined together laterally by thin bridges. The electron-microscope image is strikingly like that of the contractile axostyle of the zooflagellate *Pyrsonympha* (p. 151). Careful study shows unequivocally that each fibril arises from one of the kinetosomes along the kinety (Fig. 83, Pl. XXIII). At its base the fibril is broadly joined to the kinetosome and often appears striated, like the kinetodesmal fibrils of *Tetrahymena* and *Paramecium*. It tapers very abruptly, however, loses all trace of striation, and becomes a tubular filament of uniform diameter. All published pictures indicate that each sheet of the kinetodesmos consists thus of overlapping fibrils that enter the bundle from kinetosomes on one side and end at intervals on the other side. They clearly are much longer than the kinetodesmal fibrils of the hymenostomes. Randall and Jackson calculated that on the average there were within the bundle at any one level about 500 fibrils, while they estimated 1000 kinetosomes per kinety. This would suggest that a single fibril may extend for as much as half the length of the animal. However, there were some indications that each kinetosome might contribute more than one fibril to different sheets within the bundle.

o

PLATE XXIII

FIG. 82. Section through wall of buccal cavity of *Stentor polymorphus,* showing vacuoles and fibrillar sheets (arrows). × 30,000. From Randall and Jackson, 1958.

FIG. 83. Longitudinal section of part of one kinetodesmos of *Stentor polymorphus,* showing origin of individual fibrils at kinetosomes. × 24,000. From Randall and Jackson, 1958.

FIG. 84. Section through part of single M-band of *Stentor polymorphus*; parts of surrounding cytoplasmic vesicles are visible. × 24,500. From Randall and Jackson, 1958.

FIG. 85. Section of *Euplotes patella* showing several kinetosomes and cilia of one cirrus at upper right, with bundle of attached rootlet fibrils running toward bottom left. Note dense granules at bases and half-way up fibrils of kinetosomes; outer and inner pellicular membranes; cross-sectioned pellicular fibrils (arrows). × 24,000. From Roth, 1958b.

PLATE XXIII

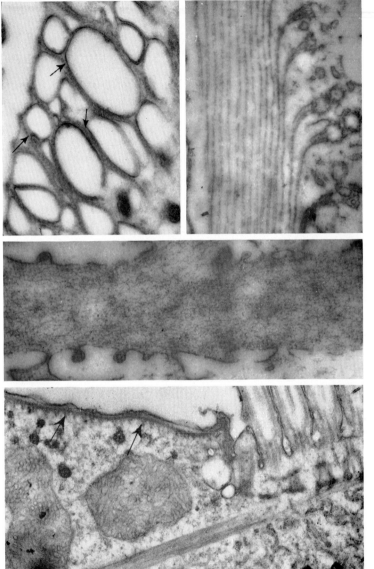

82

83

84

85

One really conspicuous way in which the *Stentor* kinetodesma differ from those of the hymenostomes and *Metaradiophrya* is that in the heterotrich the kinetodesmal fibrils pass from their kinetosomal origin to the right and *posteriad* instead of anteriad. Accordingly, the kinetodesmal bundles are heavier at the posterior end of the animal than anteriorly. At some points at least, the outer ends of some sheets within the kinetodesmos appear to be in contact with the pellicle.

It is noteworthy that in Randall and Jackson's micrograph (their Fig. 17) of a transverse cortical section cut near the adoral zone, where pellicular alveoli are present, a row of four small dense profiles appears in the cytoplasmic ridge to the right of each kinety, in precisely the place where kinetodesmal fibrils are found in the hymenostomes. This micrograph might almost have been taken of *Paramecium,* the resemblance is so striking. It seems likely that the kinetodesmal fibrils at the anterior end of the kinety start out as conventional hymenostome-type structures and aggregate in ribbons farther posteriad. A somewhat similar phenomenon occurs, it will be recalled, in *Metaradiophrya,* where posterior kinetodesmal fibrils remain free but anterior ones coalesce.

In addition to kinetodesma, *Stentor* possesses a system of longitudinal fibers running parallel to but deeper than the kinetodesma. These are called by Randall and Jackson (1958) the M-bands and by Fauré-Fremiet and Rouiller (1958a, 1958b) the endoplasmic myonemes. They are weakly birefringent and probably proteinaceous. In ultrastructure they are finely and rather irregularly filamentous, and like the endoplasmic myonemes of the peritrichs they are penetrated and surrounded by membranous vesicles and canaliculi (Fig. 84, Pl. XXIII). Filamentous strands interconnect adjacent fibers at irregular intervals and broader ones join them posteriorly. No connection to any part of the pellicle or to any ciliary organelle is reported, but in living cells they appear to be firmly inserted at the posterior holdfast region.

All observations to date, made on individuals that are at least semicontracted, indicate that both kinetodesma and M-bands are straight. According to the light-microscope studies of Randall and Jackson the kinetodesma are also straight when superextended, but somewhat crumpled in a position of minimum tension — neither contraction nor extension. The M-bands always appear straight. Clearly, both accommodate to extreme changes in length without visible folding. Randall and Jackson have considered at some length the question of contractility in *Stentor*, and we shall return to this matter in the next chapter.

The adoral membranelles of *Stentor* (Text-fig. 15) are made up individually of three rows of cilia (two rows in the buccal cavity) with 20 to 25 cilia per row. Except for the presence of fine villous extensions of the ciliary membrane there is no structure present that would serve to bind the cilia together as a unit. From the base of each kinetosome, numbers (usually about 10) of fine fibrils, about 22 mμ in diameter and apparently tubular, extend straight down toward the endoplasm. Root fibrils from the cilia of each membranelle gradually converge into a bundle extending into the endoplasm for as much as 20 μ. At their posterior extremities the bundles bifurcate and join those from neighboring membranelles, forming a zig-zag basal fiber. In addition, a thick strand of fibrous material laterally joins fibers of adjacent membranelles in the cortical region. Cross-sections of the membranelle roots cut in the region where the fibrils are converging into a bundle show a regular hexagonal pattern of fibrils linked by fine (4 mμ) filaments — an arrangement in two dimensions remarkably like the three-dimensional network in the infundibular fiber of *Campanella* (p. 195).

In view of the complexity of the total fibrillar system, and particularly the length of the overlapping kinetodesmal fibrils, *Stentor's* capacity to recover from surgical insult is truly astonishing. Tartar's (1960) recent results showing reconstitution of normal patterns after thorough mincing of the ectoplasm prove that cortical patches that have been totally disoriented have the ability to realign themselves and fuse. This follows only after the patches rotate to assume parallel or tandem, homopolar orientation, and implies in the latter case a healing or rapid regrowth of the cut ends of kinetodesmal fibrils. What is perhaps equally surprising is the persistence of a critical relationship between areas of the body having wide stripes (widely spaced kineties) and those with narrow stripes. The juxtaposition of abruptly contrasting broad- and narrow-stripe zones normally determines the site of

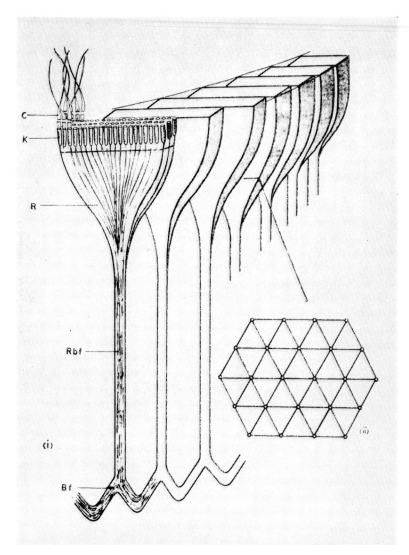

TEXT-FIGURE 15. Schematic drawing showing the organization of adoral membranelles of *Stentor*. C cilia; K kinetosomes; R root fibrils; Rbf, bundle of root fibrils from a single membranelle; Bf, zig-zag basal fiber. At right (ii) is shown the hexagonal arrangement of root fibrils and connectives as seen in cross-section at the level indicated. From Randall and Jackson, 1958.

oral anlage formation, and if these are laterally adjoined in the wrong sense (though homopolar), the induced adoral zone of membranelles will spiral in the wrong direction and produce no mouth or an unusable one. The significance of spatial relationships here — lateral distances between kineties — is harder to relate to the ultrastructural picture than is the significance of polarity and asymmetry. But *Stentor* is relatively a huge organism, and efforts to compare broad- and narrow-stripe zones in electron micrographs have not been made.

Other heterotrichs that have been examined in somewhat less detail or at less high resolution include *Spirostomum* (Finley, 1951, 1955; Randall, 1956, 1957; Pautard, 1959a, 1959b; Finley and Brown, 1960; Inaba, 1960), *Blepharisma* (Inaba, Nakamura, and Yamaguchi, 1958), *Condylostoma* (Yagiu and Shigenaka, 1958a, 1958b, 1958c, 1959a, 1959b) and *Nyctotherus* (King, Beams, Tahmisian, and Devine, 1961). *Spirostomum* and *Blepharisma* are placed in the same family and are rather similar in morphology. Both have laminated kinetodesma like those of *Stentor,* although details of the connection with kinetosomes are not available. *Spirostomum* is strongly contractile — though not so extensile as *Stentor* — while *Blepharisma* is very slightly so at best. Neither is reported to possess the endoplasmic M-bands of *Stentor*.

The pigment granules that give *Blepharisma undulans* its characteristic pink color are spheroidal bodies about $0 \cdot 5$ μ in diameter, limited by double membranes and filled with a finely granular material of moderate density. They are neatly layered immediately beneath the pellicle. A colorless strain lacks such granules but has instead abundant, smaller, very dense, irregular granules in a corresponding position.

Condylostoma has two limiting membranes over an ectoplasmic zone of packed vacuoles, some of which may be identifiable as mucigenic bodies. This zone occupies longitudinal ridges separating the furrows in which lie the kineties. Transverse furrows may also be present, but not consistently. The most striking feature of the published micrographs of *Condylostoma* is the presence of laminated fibrillar kinetodesma identical to those of *Stentor*. As in *Stentor,* the origin of individual fibrils at kinetosomes is very clear, as is their path to the right and posteriad from the kinetosome. Some pictures suggest that several fibrils arise from each basal body. At lease 14 to 16 sheets of fibrils may be counted within the kinetodesmos in some micrographs. Some species of *Condylostoma* are rather markedly contractile (Villeneuve-Brachon, 1940).

An interesting sidelight on ciliate differentiation is provided by the observations of Pautard (1959a, 1959b) on the occurrence of hydroxyapatite — the inorganic phosphate compound that is deposited on collagen during the mineralization of vertebrate bone — in *Spirostomum ambiguum*. As soil-based cultures of this species age, over a period of months, hydroxyapatite appears in small, beaded ossicles which later enlarge and fuse to form plates in organisms from very old cultures. The ossicles are seen in thin sections as a network of loosely connected islands, 1 to 3 μ in diameter, below the basal bodies, generally along the meridional lines. *Spirostomum* in these cultures, some weeks after inoculation, engages in mass burrowing activity, mining the soil (in which organic food is embedded) to a depth of centimeters. This apparently occurs after exhaustion of food in the supernatant fluid. Pautard suggests that hydroxyapatite serves a double function in these ciliates as in vertebrate bones — as an articulable skeleton providing support during tunneling activities and as a mineral bank from which phosphate can be withdrawn for metabolic needs, especially during the active "muscular" movements of tunneling.

Nyctotherus ovalis is a large heterotrich symbiotic in the cockroach gut. Its peristomal area is elongate and bears an adoral zone of membranelles subtended by a thickened ectoplasmic band. The membranelles continue into the deep peristomal tube leading to the cytostome. The ectoplasmic band extends past the cytostome and, now lacking cilia, up the opposite wall of the peristome. Somatic cilia arise singly or in pairs from more or less deep furrows. Electron micrographs show that the ectoplasm within the ridges is extensively vacuolated. Beneath, at the level of the kinetosomes, is a tangled network of randomly oriented fine filaments (about 7 mμ in diameter). The kinetosomes bear root fibrils that appear to emerge axially and pass medially; no connections with the ectoplasmic network or any other structures were observed. One micrograph shows neatly parallel tubular fibrils perpendicular to these roots, at a deeper level, but their course is unknown. Nothing resembling kinetodesma is apparent in any micrographs of body ciliature.

Primary emphasis in the paper of King *et al.* (1961) is on the membranelles and underlying structures. Each membranelle consists of four rows of 21 to 22 cilia, isolated from its neighbors by a shelf of ectoplasm. Kinetosomes within each group are directly linked by fibrils at their bases and about half-way distally; it was not determined whether these fibrils run along or across the rows of basal bodies. From the base of each kinetosome arises a group of fine fibrils that passes medially as a discrete bundle for a distance of about 1 μ, then merges with a loose hexagonal network of fibrils with dense nodes at the intersections. Root fibrils from kinetosomes of a membranelle tend to converge and show aligned nodes, so that a plate-like array of fibrils subtends each membranelle and is identified as the basal plate of light microscopists. The loose hexagonal mesh in which the basal plates are embedded constitutes the ectoplasmic band. Its nodes are about 100 mμ in diameter and the internode distance 225 to 450 mμ. On the opposite wall of the buccal cavity the meshwork continues as the posterior ectoplasmic band, but here lacks basal bodies. At certain locations structures identified as trichites occur. These consist of regularly oriented tubular fibrils, about 20 mμ in diameter. In some micrographs these appear to replace the basal plates, in others they are clearly connected with the ecto-plasmic band but their relationship with kinetosomes is not shown. Some micrographs suggest the presence of bundles of fine filaments forming a karyophore around the nucleus. In many respects the ectoplasmic band structure of *Nyctotherus* resembles the reticulate infundibular fiber of the peritrich *Campanella,* although the network appears more precise in the latter instance. The converging root fibrils of each membranelle also roughly resemble the membranellar fibrils of *Stentor.*

Order Entodiniomorphida

The Order Entodiniomorphida represents one of the peaks of ciliate evolution. In these spirotrichs, all endocommensal in the digestive tracts of herbivorous mammals, the ciliature is much reduced. Some genera retain only an adoral zone of membranelles that function both in feeding and in locomotion; most forms have

PLATE XXIV

FIG. 86. Transverse section into cortex of ophryoscolecid ciliate, in region where layers of pellicle are thin. Outer membrane covers homogeneous second layer; below this, longitudinal fibrils are seen as orderly bundles of circular profiles; innermost layer is composed of transversely oriented filaments. × 32,000. From Noirot-Timothée, 1960.

FIG. 87. Section of ophryoscolecid ciliate showing kinetosomes of adoral ciliated band with fine interconnecting or oblique fibrils; massive, dense, banded basal rods; and retrociliary fibers composed of regularly packed fibrils. × 32,000. From Noirot-Timothée, 1960.

FIG. 88. Section through anterior end of esophagus of ophryoscolecid ciliate. Deeply folded esophageal wall is scalloped line along left and lower right sides of picture. Cavities at upper right are sections through lumen of peristomal canal, subtended by fibrillar sheets. Retrociliary fibers, composed of packed fibrils, are cross-sectioned above and within esophagus. × 21,000. From Noirot-Timothée, 1960.

PLATE XXIV

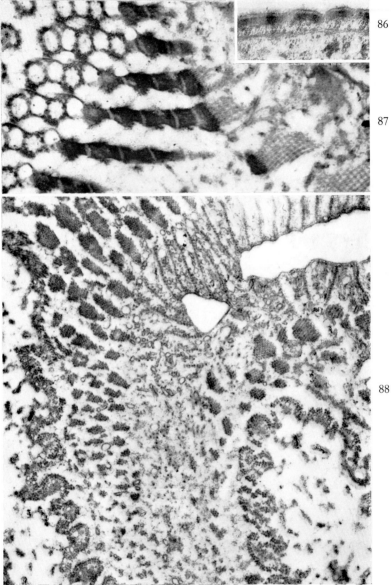

86

87

88

in addition a dorsal zone of membranelles, and in some, posterior rows or tufts of membranelles occur. A cortical layer of cytoplasm encloses the nuclei, contractile vacuoles, and usually one or several skeletal plates. Most of the body volume is occupied by an endoplasmic sac, sharply demarcated from the cortical zone in stained specimens, and leading via a tube called the esophagus to the mouth and via a rectum to the anal pore. The body is relatively rigid but the anterior expanded buccal cavity or peristome, with its encircling adoral ciliature, can be retracted and covered over by surrounding cytoplasmic lips.

Several genera of the Family Ophryoscolecidae have been extensively studied by both light and electron microscopy by Bretschneider (1959, 1960) and Noirot-Timothée (1957, 1958a, 1960). The two authors are in general agreement on details of ultrastructure but their interpretations of these data relative to the light-microscope picture diverge somewhat — a fact explainable not by any lack of scrupulous consideration on their part but by the fantastic complexity of ophryoscolecid morphology. The species they have considered, all from the rumina of cattle and sheep, are not always distinguishable in electron micrographs, and seem sufficiently similar in ultrastructure to be described together.

At the surface of the ophryoscolecid body is a highly differentiated cortical zone consisting of at least four layers (Fig. 86, Pl. XXIV). (1) The outermost is apparently a unit membrane, and often appears to bear a fine granulation or fuzz on its outer surface. Beneath this is (2) a homogeneous, moderately dense layer, irregularly traversed by low-density canals. Its thickness varies widely with species and with body region, since its outer border may be thrown into folds, regular ridges, or knobs. It apparently is composed of a glycoprotein. Its inner surface is relatively smooth or minutely furrowed to accommodate (3) an underlying system of orderly, parallel, longitudinal fibers. Each of these consists of 15-mμ tubular fibrils arranged in two rows of two to four each, the bundle assuming a regular rectangular shape in cross-section. Below these longitudinal fibers is (4) a layer that varies in thickness from nothing to well over 1 μ. It seems to be composed of a finely filamentous material arranged generally transversely and interrupted at intervals by transversely oriented fissures.

The adoral ciliature consists of a long spiral band in which cilia occur in short oblique rows, each row arising from a basal rod visible in the light microscope (Text-fig. 16). Electron micrographs of sections through the level of the kinetosomes show that the rows are uniformly aligned, and not separated into groups corresponding to discrete membranelles. Each kineto-

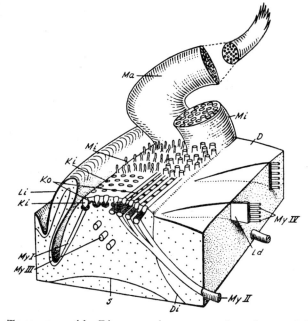

TEXT-FIGURE 16. Diagrammatic reconstruction of part of the adoral ciliary zone of an ophryoscolecid ciliate. Two syncilia are shown at the rear, labeled Ma and Mi. In front of these are shown, on the right, the stubs of cilia making up another syncilium and, on the left, microvilli labeled Mi, that protrude from the cell membrane between cilia. In front of this are the apices of kinetosomes (Ki) on the left and the outlines of basal rods on the right. Noirot-Timothée has shown that in at least some species the basal rods pass obliquely across the band rather than longitudinally as shown here. Commissural fibers (Ko) are shown connecting the kinetosomes laterally. On the cut faces of the stereogram are seen various fibrous structures considered by Bretschneider to be myonemes. My I, II, and III are retrociliary fibers passing in different directions. My IV and Ld together are the fibrillar sheets subtending the peristomial disc (D). From Bretschneider, 1960.

some, topped by a cupped diaphragm and a distinct axial granule, typically contains in its lumen a single dense granule. It is closed proximally by an intimate junction with the basal rod of its row. The rod is a massive, dense structure, somewhat greater in diameter than the kinetosomes themselves, and marked by rather regular variations in diameter and in opacity (Fig. 87, Pl. XXIV). In spite of the continuously oriented aspect of the infraciliature, the cilia in action group themselves into discrete clumps. According to Noirot-Timothée these should be called syncilia rather than membranelles, since their cohesion is loose and is not reflected in the infraciliature. In thin sections cut above the level of the cell surface they appear as well-ordered groups of six to nine rows of about 15 synchronized cilia each, arranged as a double tier of syncilia along the spiral band. This means that cilia arising from a single oblique row (subtended by a single continuous basal rod) contribute to at least two syncilia.

In addition to their oblique linkage via the basal rods, kinetosomes are connected to each other along, and probably between, rows by lateral bridges. Furthermore, slender dense fibrils arise from one border of each basal body and run past at least the next kinetosome in the row. These are identified as probable kinetodesmal fibrils by Noirot-Timothée.

From the basal rods beneath the kinetosomes arise series of retrociliary fibers of very specific structure (Figs. 87 and 88, Pl. XXIV). Like so many other intracellular fibrous bodies, they consist of numbers of tubular fibrils, about 15 mμ in diameter, arranged in orderly parallel planes. They are encountered in sections at many locations within the anterior ectoplasm, and it is not possible to determine their course from electron micrographs. Light-microscope studies show that they vary considerably in length and that groups of them pass toward and perhaps into the esophagus or along its ectoplasmic surface, toward the base of the cytoplasmic lips surrounding the ciliated zones, toward the skeletal plates, and perhaps in a spiral beneath and parallel to the ciliated bands.

Over the surface of the peristomal disc within the adoral ciliated zone and down its canal to the cytostome, the pellicle lacks the three inner layers described for the general body surface. Instead there exists a series of regularly spaced fibrous sheets lying normal to the cytoplasmic surface (Fig. 88, Pl. XXIV). Each consists of two layers of tubular 15-mμ fibrils oriented perpendicular to each other. Minute clear vesicles typically are abundant in the cytoplasm between these sheets.

The esophagus, according to Noirot-Timothée, is a funnel, open anteriorly, into which the peristomal canal, terminating in the cytostome, depends eccentrically. The esophagus wall consists of an outer, finely filamentous layer and an inner array of 15-mμ tubular fibrils, grouped as two-layered packets of 10 to 12. Through the open anterior end of the esophageal funnel there is continuity of ectoplasm and endoplasm, at least intermittently. However, the fibrillar elements of the esophagus wall may perhaps be continuous in some regions with either the fibrils of the subperistomal sheets described above, or with retrociliary fibers that descend in this direction. Along its dorsal surface the esophagus wall is thrown into very straight, narrow, deep folds protruding into its lumen (Fig. 88, Pl. XXIV); elsewhere it is irregularly undulating. About midway down the body, the esophagus spreads out over the dorsal surface of the endoplasmic sac and merges with the latter. The wall of the endoplasmic sac, which appears distinctly in the light microscope as a membrane or fibrous sheet, is less discrete in most electron micrographs. In some areas it seems to be a thinned-out version of the esophagus wall; elsewhere it consists of an ill-defined series of filamentous cords. It apparently is not a continuous layer, and its only membranous component is an array of granular endoplasmic reticulum flattened against its inner surface.

Endoplasm and ectoplasm are similar in structure, except that Golgi bodies are limited to the endoplasm. The existence of typical mitochondria is questionable; Noirot-Timothée believes that membrane-limited spheroids with ill-defined internal contents may represent quasi-mitochondria in these anaerobic cells.

At the posterior end of the cell, the rectum is visible as a distinct inpocketing of the pellicle. Where it meets the boundary of the endoplasmic sac, the filamentous internal layer of the pellicle seems to be continuous with the filamentous component of the sac wall. The rectum itself is completely open to the exterior; where it abuts against the endoplasm the latter is bounded by a single unit membrane.

In addition to the retrociliary fibers coursing every which way through the anterior ectoplasm, tracts of filamentous material that apparently derive from the innermost layer of the pellicle are seen in electron micrographs of the region of the rectum and at various sites in the anterior part of the body. A network of such filaments spreads over the outer surface of the skeletal plates and extends anteriad as a heavy tract to a sizable aggregate of inter-laced filaments in the ectoplasm adjacent to the adoral ciliated zone. Noirot-Timothée identifies this aggregate as the so-called motorium of light microscopy. Some other bands of filamentous material she concludes represent caudal, rectal, peri-esophageal, and circumciliary fibrils seen in hematoxylin-stained light-microscope preparations. Electron micrographs did not permit recognition of a complex silverline system concentrated around the ciliated zones.

The skeletal plates of the ophryoscolecids consist of packed prisms of homogeneous, low-density material, lacking limiting membranes but often surrounded by aligned cisternae of granular membranes. They are composed of a neutral polysaccharide, not glycogen. Noirot-Timothée considers that they represent nutritive reserves, perhaps secondarily providing mechanical support. Other polysaccharide granules are commonly distributed throughout the cytoplasm in well-fed organisms.

Both Bretschneider and Noirot-Timothée note the occurrence, on one side of the adoral zone of syncilia, of a single row or small clump of cilia that are not aligned with the adoral band and that in cross-section look quite different. Their diameter is greater than usual owing to an expanded matrix zone between the axial fibril bundle and the surface membrane, and the matrix is notably less dense than that of ordinary cilia. Bretschneider suggests that these may be specialized sensory organelles.

A particularly intriguing question concerns the ontogeny of the ciliated zones in the entodiniomorphs. The first evidence of impending cell division is the appearance of an adoral zone anlage within a vesicle in the ectoplasm of the presumptive posterior daughter cell, far removed from the parent ciliature. Noirot-Timothée shows electron micrographs of several of these vesicles, which are lined by a unit membrane and contain rows of kineto-somes along the inner wall. In later stages cilia emerge from the kinetosomes into the empty vesicle. The kinetosomes are of conventional appearance and seem to be interconnected by fibrils of uncertain disposition, but lack the central granules of kinetosomes in the adult ciliature. The origin of these kinetosomes is in doubt. Some authors (see Noirot-Timothée, 1960) consider that a somatic infraciliature is present in the unciliated regions of the adult body. Noirot-Timothée finds no such system in her light-microscope preparations but has encountered in a very few instances scattered kinetosomes beneath the body surface in thin sections. Whether these are persistent somatic kinetosomes or new-formed rudiments of an adoral zone anlage could not be determined. The question is thus left wide open.

An interpretation of the bewildering assortment of fibrous and filamentous structures encountered in these extraordinary ciliates is at best highly speculative and teleological. As noted above, the body form is quite constant, only the ciliated zones and surrounding lips being notably contractile. During feeding, large plant fragments may be ingested; this requires active movement of the lips and may even result in some body distortion. Bretschneider considers that the retrociliary fibers must necessarily serve as myonemes, since their abundance and distribution would seem to qualify them for this role. In his view, the entire four-layered cortex provides mechanical support and by its slightly elastic properties assures recovery from distortion and causes eversion of the retracted ciliated region when the myonemes relax. Noirot-Timothée recognizes that the retrociliary fibers may be contractile but also considers the possibility that the innermost filamentous layer of the cell surface and the tracts of similar filaments coursing through the body may be the myonemes, the retrociliary fibers in this event serving as elastic antagonists.

Order Hypotrichida

The final group of ciliate protozoa is represented in the electron microscope literature by its most familiar member, *Euplotes,* the object of a detailed study by Roth (1956, 1957). Additional observations on nuclear structures of *Euplotes* and *Stylonychia* have been recorded in Chapter 2.

In the hypotrichs, somatic ciliature is reduced to a small number of heavy cirri (tufts of coherent cilia), typically arranged in precise groupings on the ventral surface of a flattened, semi-rigid cell, and used as legs in the ciliate's characteristic crawling locomotion. The adoral band of membranelles is well developed and functions importantly in swimming. Light microscopists have described intracellular "neuromotor" fibers interconnecting the cirri, the membranelles, and a motorium, and a net-like silverline system occurs in constant and species-specific patterns (see recent revision of the genus *Euplotes* by Tuffrau, 1960).

The pellicle of *Euplotes patella* (Fig. 85, Pl. XXIII) consists of at least two unit membranes, the outer one being continuous over the surfaces of the cilia and the inner one forming a thickened ring about the cilium bases. Beneath the inner layer are two mutually perpendicular systems of fine tubular fibrils about 22 mμ in diameter. Cirri arise from depressions in the cell surface and are composed of groups of about 27 cilia arranged with great precision in five to eight parallel rows. Frequent tubular out-pocketings of the ciliary membrane are the only structures seen that could, if they intertwine, account for the adhesion of cilia within a cirrus or a membranelle. Internally, the kinetosomal fibers end at dense granular thickenings, and similar granules occur around the kinetosomes in a plane about half-way distally. At both these levels, fibrils may interconnect adjacent kinetosomes directly or may pass them tangentially in a manner rather suggest-ing the orientation of kinetodesmal fibrils.

The membranelles are similar in structure to the cirri, except that each membranelle consists of two rows of 15 to 25 cilia, occasionally with an additional short row. From the proximal ends of some kinetosomes of both cirri and membranelles arise rootlet fibrils — several fibrils per kinetosome but not from every kinetosome in the group. These are, again, tubular 21-mμ fibrils. They may pass vertically into the cytoplasm, laterally in bundles toward other ciliary organelles, or peripherally toward the pellicular fibril systems. In the region of the cytostome appears an aggregation of fibrils running in several directions within the mass and extending from it at least toward the buccal cavity wall and toward neighboring membranelles. Roth identifies this mass with the light microscopist's motorium and the bundles of tubular fibrils with their neuromotor fibers. In addition, ill-defined tracts of similar fibrils were seen adjacent to the micronucleus.

The dorsal bristles of *Euplotes* are cilia arising singly or in pairs from conical depressions of the body surface. Cilium and kinetosome structure are typical, but no associated fibrous structures are discernible. In light-microscope preparations, six or more rod-shaped granules appear, surrounding each bristle. Electron micrographs reveal these to be vacuoles bounded by a single membrane and containing a homogeneous material of low density. Similar vacuoles are observed singly or in small clusters near the bases of the cirri and membranelles. The dorsal bristles are supposed to be non-motile and have been suspected of serving a sensory function.

No structures were observed that would account for the silverline system.

CHAPTER 7

CONCLUSIONS

In the foregoing pages we have surveyed, in a necessarily cursory manner, a formidable body of information on protozoan ultrastructure. Yet one need but glance at any textbook of protozoology to appreciate how small a sample has been selected from the legions of existing species. And for very few of those studied is the available information anything like complete — as the workers who have investigated them are the first to point out. How can we presume to draw conclusions from data that are fragmentary, scattered, and based almost exclusively on observations of micromorphologic appearance unsubstantiated by evidence from any other quarter? It must be admitted — in fact, insisted — that the effort is presumptuous and that it can consist only of a summing-up of impressions, inferences, and intuitions.

Impressions concerning organelles with a wide distribution in nature seem the least precarious and have already been presented in Chapter 2. For the most part they do not need reiteration. Other speculative considerations have been sprinkled through the discussions of various groups. In this chapter are considered, first, some of the novel structures repeatedly encountered in protozoa; perhaps their distribution offers clues to their significance. Then we may inquire whether the facts of ultrastructure provide any new insights into relationships among protozoa and other organisms.

Membrane Differentiations

Almost all protozoa examined prove to have a unit membrane at the actual cytoplasmic surface. Possible exceptions include *Gromia,* where a highly organized lamella is found immediately under the shell, and no conventional plasma membrane is apparent in the material studied. In heliozoa and filose rhizopods the existence of a semipermanent membrane over adjacent antagonistic streams of cytoplasm seems mechanically implausible, and the electron-microscope picture needs further investigation. All theories of ameboid movement require either a fairly free slippage of the plasmalemma over the ectoplasm or a rapid interconversion of membrane and superficial ectoplasm. The former seems more likely, although in a great many protozoa the probability that membranes can be synthesized very rapidly finds strong support in electron-microscope data.

Amebae and most flagellates appear to be limited by a single unit membrane. For the sporozoa, euglenoids, and ciliates, two or more membranes are commonly described. There is something conceptually disturbing about the idea of two limiting membranes. What does the outer one limit? Or if, as is usually reported, the outer is the only one that is continuous over cilia or potential orifices, then what becomes of the inner one here? A permanently discontinuous membrane is hard to reconcile with the usual picture of membrane formation and function; cells and membranous organelles generally consist of closed sacs, vesicles, and tubes. If two true membranes are present, one would expect that something is present between them that must be limited on both sides, and that both membranes are continuous, defining a completely enclosed peripheral space. This may be true for *Plasmodium,* for example, but the situation in euglenoids and many ciliates remains enigmatic.

For at least some ciliates, the presence of three superficial membranes is established, the two inner ones composing the mosaic of enclosed alveoli that define the silverline pattern. This has now been demonstrated in *Paramecium,* in three tetrahymenids, and in one astome, and the question arises whether a similar system exists in other ciliates. Two difficulties prohibit an easy answer. In the first place, *the* silverline system is not identifiable because different methods of silver impregnation may affect different morphological entities. The commonest (Klein, Gelei, and Chatton-Lwoff) techniques reveal kinetosomes and superficial linear patterns in most, but not all, ciliates (as well as in many flagellates, sporozoa, etc.); other methods may impregnate internal structures. In the second place, most electron microscopists have failed to observe any structures corresponding to the silverline patterns described for their subjects. Precise discrimination in electron micrographs of multiple surface membranes requires a considerable measure of luck as well as high resolution, and is easier if pellicle fragments have been examined as well as sections; it is hardly surprising that investigators not specifically looking for this sort of evidence have given ambiguous reports.

But the classical ciliate silverline system is not a mirage. The alveoli of the hymenostomes make morphological sense, and the most reasonable supposition at present is that a like situation will be found to obtain in other forms with equally conspicuous patterns demonstrated by the same methods.

At the moment, the only function known for the silverline system is to aid the protozoologist in identification and developmental studies of his cells, but presumably the ciliate has some less altruistic reason for preserving it. The pellicular alveoli spread out between meridional rows of organelles — cilia and trichocysts or mucigenic bodies — emerging from the cell surface. In *Paramecium* they are transversely dissected into orderly compartments corresponding to ciliary units; in the tetrahymenids they extend longitudinally past several to many ciliary units along each kinety, transverse interruptions being apparently random and variable in time. Their general distribution clearly is dictated by the disposition of kineties. Yet something else is involved in the specific silverline pattern, for in many species it assumes the form of a network in areas between kineties, or shows highly characteristic configurations in parts of the body that lack ciliature and infraciliature. Specific variations conceivably may be as incidental and non-adaptive as are some color variants in other organisms, but the alveolar system itself must have a highly significant adaptive value.

The alveoli may contribute to the rigidity of the pellicle. *Paramecium* and the tetrahymenids upon appropriate treatment readily yield ghosts or extensive surface fragments, attesting to a considerable structural integrity in the pellicle. *Stentor* and *Spirostomum,* by contrast, do not; their surfaces have to be highly deformable to permit the extreme contraction or extension of the cell. Randall and Jackson (1958) showed that very minutely dissected alveoli or strings of vesicles are present under the superficial unit membrane in some parts of *Stentor;* typical *Paramecium*-like alveoli occur only near the adoral zone.

Villeneuve-Brachon (1940) reported that *Stentor, Spirostomum,* and *Blepharisma* (which is non-contractile) lack any silverline system, but that *Condylostoma* (contractile) has a well-developed linear pattern with transverse connections. The implications are paradoxical. It would appear that the minute vesicular structures of *Stentor* do not bind silver, and that rigidity cannot be a consistent property of all silverline systems.

The alveoli must inevitably affect membrane activities at the cell surface. Their response in living cells to variations in osmotic pressure has not been investigated. They are present over most of the buccal cavity of the tetrahymenids, but the actual cytostome has not been studied at high resolution. They appear to be absent in the buccal cavity of *Stentor,* where a rapid uptake of water is presumed to occur; instead, the whole cortical cytoplasm is filled with vesicles. Whether alveoli may somehow be involved with membrane transmission of excitation is a totally new question.

For the intracytoplasmic membrane components of the protozoa, there is nothing significant to add to the discussion in Chapter 2. Closer looks are needed at both the structural and the biochemical properties of many bodies identified as mitochondria in anaerobic organisms. Presently unclassifiable structures — such as the clouds of tiny vesicles described in *Opalina* (p. 167) or whorls cf concentric membranes — are not uncommonly encountered in protozoa, as in other cells. Some of them perhaps are related to the cell's means for getting rid of used furniture and excess baggage, or they may be stages in the development of other organelles. Membranous structures often appear to be specifically associated with fibers or filaments, as is noted below.

Fibrous Structures

If there is anything unique in the ultrastructural organization of protozoa, it may be the extent to which they have utilized fibrous materials in the construction of cytoplasmic organelles and organelle systems. A bewildering profusion of fibrous structures has been described, including some in each protozoan class.

To begin with, fibrous elements can be listed according to their most obvious, and least reliable, property: what they look like individually. Some fibrillar structures for which available evidence does not permit definite classification are included in parentheses under categories to which the author guesses they belong. Asterisks refer to kinetosomal attachment, as will be explained below. Exclusively intranuclear fibers are not included.

1. *Banded fibrils, exhibiting a period of 30 to 60 mμ and varying greatly in width; possibly made up of bundles of periodic filaments in phase.*

> *Rhizoplasts or other flagellar root fibrils of phytoflagellates: *Chromulina psammobia, Synura caroliniana, Paraphysomonas vestita, (Chrysochromulina ericina), (Micromonas squamata), Pedinomonas tuberculata.*
>
> *Costa of *Trichomonas, Tritrichomonas.*
>
> *Parabasal fibrils of *Tritrichomonas, Trichomonas, Foaina, Trichonympha.*
>
> *Axial ribbons of *Pyrsonympha, Joenia.*
>
> *Kinetodesma of *Paramecium, Tetrahymena, Colpidium, Glaucoma, (Ophryidium), Metaradiophrya, (Stentor), (Opalina).*
>
> *Stalk fibrils of *Opercularia, Carchesium, Zoothamnium.*
>
> Discharged trichocysts of many holotrich ciliates and of the flagellate *Oxhyrris marina.*
>
> Conceivably, the accessory ribbons of the flagella of euglenoids.

2. *Cylindrical fibrils described as tubular, with dense peripheries and low-density centers; ranging in measured diameter from 12 to 30 mμ.*

> a. Fibrils at least slightly sinuous, never in definite sheets or lattices.
>
> > *Spindle fibers.
> >
> > Fibrils in mucopolysaccharide capsule of *Gromia.*
> >
> > Fibrils in axopodia of *Actinophrys.*
> >
> > Fibrils in external cyst wall of *Sarcocystis tenella.*
>
> b. Fibrils straight or nearly so; regularly spaced or arranged in conspicuously orderly arrays.
>
> > *All kinetosomes, cilia, and flagella.
> >
> > *(Fibrils in pellicle of *Tetramitus*).
> >
> > Minute cylinders in lamellae beneath shell of *Gromia.*
> >
> > Fibrils in extremities of pellicular ribs of gregarines.
> >
> > Longitudinal pellicular fibrils of sporozoites of *Plasmodium, (Sarcocystis), (Toxoplasma).*
> >
> > (Fibrils in polar filament of *Nosema*).
> >
> > Fibrils in haptonemata of *Chrysochromulina.*
> >
> > *Pellicular fibrils and kinetosomal connectives in *Euglena, Peranema.*

Rodorgan of *Peranema.*
Pellicular and "cytostomal" tube fibrils of trypanosomes, *Cryptobia, *Bodo.*
Axostyle walls of *Tritrichomonas, Foaina, Joenia, Lophomonas, Holomastigotoides.*
Entire axostyle of *Pyrsonympha.*
Root of attachment organelle of *Pyrsonympha.*
Rostral cap pellicular fibrils of *Trichonympha.*
Pellicular fibrils of *Opalina.*
Longitudinal, *transverse and *postciliary fibrils of *Tetrahymena, Colpidium, Glaucoma.*
Fibrils of contractile vacuole wall of *Paramecium, Tokophrya, Metaradiophrya.*
Trichites of gymnostomes, *Frontonia, Nyctotherus.*
Fibrils in sucking tentacles of *Ephelota,* (*Discophrya*).
(Myonemes of *Metaradiophrya*).
*Stalks of *Opercularia, Epistylis,* (*Vorticella*).
*Kinetodesmal fibrils of *Stentor, Spirostomum, Blepharisma, Condylostoma.*
*Membranelle roots of *Stentor.*
Fibrillar sheets subtending buccal cavity of *Stentor.*
Longitudinal pellicular fibrils, fibrils of esophagus wall, *retrociliary fibers, fibrillar sheets subtending buccal surface in ophryoscolecids.
*Pellicular and ciliary root fibrils of *Euplotes.*

3. *Fine filaments, less than 15 mμ in diameter, often in bundles or sheets.*
Axopodia of *Actinosphaerium.*
Circular myonemes and smooth cortical layer of gregarines.
Ectoplasmic network of *Gregarina rigida.*
(Paraxostyle of *Pyrsonympha*).
Infraciliary lattice of *Paramecium.*
*Network under buccal cavity membrane in *Paramecium* and tetrahymenids.
Fibers in prehensile tentacles of *Ephelota.*
Ecto-endoplasmic boundary and karyophore of *Isotricha.*
*Ecto-endoplasmic boundary layers of *Metaradiophrya.*
*Some kinetosomal connectives in *Ophrydium.*
Stalk and *endoplasmic body myonemes of peritrichs.
*Filaments in infundibular fiber of *Campanella.*

Endoplasmic myonemes or M-bands of *Stentor;* transverse
  bands connecting membranelle roots distally.
Ectoplasmic network and *ectoplasmic band of *Nyctotherus.*
Innermost cortical layer, esophagus and endoplasmic-sac walls,
  filamentous tracts in various locations in ophryoscolecids.
(Tracts and meshworks of filaments near buccal cavity of
  Euplotes; these perhaps belong under 2a).

In addition to all of these there are several examples of probable
fibrous connections that are unclassifiable. These include prominently the short bridges that interconnect kinetosomes, often at
two specific levels, in *Lophomonas, Nyctotherus,* ophryoscolecids,
Euplotes, and some other ciliates.
Now it is clear that in appearance alone one cannot find a
consistent fundamental meaning, even if one allows for descriptive
errors resulting from imperfect fixation or low resolution.
Whereas most of the fibers whose chemical composition has been
examined at all appear to be mainly protein, the oral capsule of
Gromia is mucopolysaccharide; the fibrils named in category 2a
probably have nothing in common. Furthermore, the categories
are not mutually exclusive. The ciliary fibrils of the scopulae of
peritrichs start out as distinct tubules and may become banded
fibrils. The kinetodesmal fibrils of *Stentor* look striated near their
origin but become distinct tubules upon joining the kinetodesmos.
One more property that certainly has fundamental significance
may be added to the above listing: fibrils that are known to
originate at or to be directly connected with kinetosomes or
centrioles are marked with an asterisk (*). Some others, unmarked
in the list, are indirectly connected or occur in such positions that
connection with kinetosomes/centrioles seems a strong possibility.
  It appears that all of the intracellular fibrils displaying a banded
appearance originate at kinetosomes or centrioles. The point of
attachment is typically lateral, near the base, but may be terminal.
In ciliates thus far studied, the banded kinetodesmal fibril
invariably passes parallel to the cell surface, to the right, and
anteriad or (only in *Stentor* and related heterotrichs) posteriad. In
flagellates the fibril may pass in any direction from its origin — up
toward the cell surface, laterally toward chloroplasts, etc., or
medially to and past the nuclear surface. Similarly banded fibers

are very common in ciliated cells of metazoa; they frequently originate at the base of a tapering, closed kinetosome and usually pass medially, resembling the rhizoplasts of flagellates rather than the kinetodesma of ciliates.

The only fibrils that appear unquestionably to be direct outgrowths of kinetosomal fibrils are those of the flagellum itself, sometimes converted into something else as in the stalks of peritrichs.

Among the tubular and filamentous fibrils, a large proportion is known to be, or suspected of being, connected with kinetosomes, but there remain some for which such a suspicion seems tenuous, although it cannot be eliminated. Conceivably a centriole could be so located as to participate in the organization of pellicular fibrils in the sporozoans, for example, and the pellicular fibrils of *Opalina* might originate at kinetosomes of the falx area, which has not been determinedly studied. But the necessity or permanence of this relationship should not be exaggerated. In *Euglena*, for example, where a relatively small number of fibrils takes off from the kinetosomes to join the pellicular system in the reservoir, the system over the body surface embraces a large number of parallel fibrils, and some of the surface striae do not enter the reservoir at all. In several zooflagellates, the axostyle seems not to connect directly with the centriole/kinetosome complex but rather to be linked to it by fibrils of a different kind. New axostyles do, however, grow out from the centrosomal region. In *Tetrahymena*, the longitudinal fibril bands clearly have no direct connection with kinetosomes. In the ophryoscolecids, filamentous layers occur in profusion all over a body that has a relatively small ciliated zone. An aspect of the problem of fibril morphogenesis might be approached by studying *Tetramitus*, in which an ameboid cell may transform into a flagellated one (a process probably involving the appearance of pellicular fibrils as well as of flagella) within less than an hour.

The multiplicity of kinds (appearances!) of fibrils arising in the kinetosome region poses many unanswerable questions. A tetrahymenid somatic kinetosome somehow "knows" that it has to organize one kind of fibril at its right anterior margin and another kind at its right posterior margin and left side; perhaps these events occur at different times when different precursors are present to be organized; or perhaps they are made from the same original materials, but asymmetric physical or chemical fields in the surrounding cytoplasm determine how they will be linked into fibrils and which direction they will take. The problem is so remote from present understanding that more speculation seems useless.

Given a plethora of fibrillar designs and constructions, we can scarcely conceive of enough functions to go around. A few come obviously to mind: contraction, mechanical support, information transfer.

As previously mentioned in this review, currently respected hypotheses of the mechanism of protoplasmic contraction require that this occur either by the folding of protein chains or else by the sliding of interdigitating filaments (actin and myosin in the case of striated muscle) within a closely packed bundle. Contraction by folding of chains in a gel network is adduced to explain ameboid locomotion according to more classical theory, but both mechanisms or something like the second one alone could operate in ameboid and foraminiferan movement according to recently proposed views.

Morphologically identified fibrous structures that certainly possess the property of contractility are flagella and cilia, which indubitably beat, and the myonemes of vorticellid stalks, which unquestionably shorten. We have nothing here to add to the discussion of the former in Chapter 2. The myoneme of the vorticellid stalk consists of a long bundle of roughly parallel fine filaments of indeterminate length, penetrated and incompletely surrounded by membranous canaliculi. Its general similarity to smooth muscle (which has been relatively neglected by biophysicists) has been repeatedly pointed out.

If the filamentous vorticellid myoneme be accepted as a truly contractile organelle, then it becomes reasonable to assume that similar constructions also have this capacity. These include the endoplasmic body myonemes of peritrichs, the M-bands of *Stentor,* and the axes of prehensile tentacles in *Ephelota*. Of the other structures in class 3 above, the central filament bundle in *Actinosphaerium* axopodia; the circular myonemes of gregarines; the layered, filamentous, ecto-endoplasmic boundary in the astomes and *Isotricha;* and the interconnected, pellicular and digestive-sac filamentous layers of the ophryoscolecids seem possible candidates for the distinction of being contractile. Most of these are continuous with tracts or bundles that depart from the layers, and in several instances aggregations of tiny membranous bodies are described as following the contours of the filamentous structures.

Of all of these filamentous organelles, only the ecto-endoplasmic boundary of astomes and some of the body myonemes of peritrichs have been described as directly and consistently associated with kinetosomes. Oddly enough, for several of the others, no insertion on any other body component has yet been detected.

The conclusion that some of these filamentous organelles are contractile seems logical and inevitable, but just as inevitably it must be recognized that these cannot be the only organelles specialized for contraction in the protozoa. Unless secondary filaments or the matrix in flagella are the exclusive sites of molecular shortening or sliding, the tubular fibrils must be involved. Other organelles composed of tubular fibrils are equally suspect.

We may consider first the well-documented morphology of *Stentor* (see also the thoughtful discussion in the paper by Randall and Jackson, 1958). Observers of living and stained cells in the light microscope have identified as myonemes the structures recognized by electron microscopy as kinetodesma. Both M-bands and kinetodesma, as stated earlier, remain straight under conditions of body contraction and maximal extension. Slippage against each other of the overlapping fibrils in each row, firmly attached to kinetosomes at one end and individually free on the opposite side of the kinetodesmos at the other, could result in very significant shortening and lengthening of the whole fiber, although the availability to each fibril of only two others to hang on to makes the situation rather different from that in striated muscle. Perhaps filaments of two different kinds are essential for the slipping-zipper model, but since we know nothing whatever of the chemical composition of kinetodesma, we can conveniently shrug off this difficulty for now.

In *Spirostomum* and *Condylostoma* — both contractile — only kinetodesma, and no M-bands, have been reported to date, but more detailed study of both genera is needed. The suggestion of Randall and Jackson that the two systems (filamentous and tubular) may complement and reinforce each other to accomplish the extraordinarily dramatic changes of form in *Stentor* is attractive. But *Blepharisma* has similar kinetodesma. Perhaps *Blepharisma* is contractile and doesn't know it — no less silly explanation suggests itself.

The sucking tentacles of *Ephelota* are said to be extensile and contractile. They lack the filamentous axis of the prehensile tentacles and have instead an orderly arrangement of tubular fibrils in sheets and crests. The haptonema of *Chrysochromulina* are, without question, contractile; no structure other than membranes and tubular fibrils has been discerned in them. The accessory fibers of the *Peranema* rodorgan, strongly suspected of moving that body, are sheets that appear fibrillar. The contractile axostyle of *Pyrsonympha* consists exclusively of tubular fibrils resembling in arrangement the kinetodesmos of *Stentor* (but the significance of the adjacent, filamentous paraxostyle is unknown). Only tubular fibrils have been found associated with ciliate contractile vacuoles. Other functions might be suggested for these fibrils, but *something* has to hold injector and ejector canals closed while ampullae or vacuoles are filling; simple elasticity can hardly suffice. Retrociliary fibers of ophryoscolecids are reasonably suspected of being contractile (for organisms with relatively rigid bodies, they are remarkably well supplied with possible contractile systems, like protozoan arthropods).

None of the striated fibrils can be seriously suspected of being actively contractile on present evidence — even though *Stentor* kinetodesma appear to arise as banded fibrils. Most of them fit agreeably enough into a provisional category labeled "elastic supporting and anchoring structures".

Many of the tubular-fibril bodies certainly do not themselves contract. Often these are composed of fibrils in precise geometric array — unlike the slightly curved laminae of the *Pyrsonympha* axostyle and the heterotrich kinetodesmos. Such are the skeletal protein fibers of the gymnostome pharyngeal basket, the rods of the *Peranema* rodorgan, the membranelle roots of *Stentor*. The retrociliary fibers of ophryoscolecids are likewise precisely packed, and this might constitute an argument against the assumption that they are contractile, were not actin and myosin filaments in some muscles hexagonally packed.

The peripheral fibrils of *Tetramitus,* euglenoids, trypanosomids, and bodonids deserve mention. These organisms are slightly to extremely plastic; a supportive function for the fibrils is strongly indicated, although there is room for suspicion of contractility. Pellicular fibrils of *Trichonympha* (rostral cap), the tetrahymenids, and *Euplotes* may also provide mechanical strength. Possibly the fibrillar sheets subtending the buccal cavity membranes in *Stentor,* the ophryoscolecids and, according to Noirot-Timothée (1960), several other ciliates, do too.

These lists could be extended by considering inconclusive evidence about many other fibrillar organelles, but in view of the magnitude of uncertainty about even the best-known examples, this speculation has gone far enough.

The third possible function proposed above was information transfer, surely the most controversial of all — we are descending from attractive probability through extreme uncertainty to total confusion. Presumably any structure with distinct spatial organization that is capable of transferring electrons — as protein fibers are, according to Szent-Györgyi (1960) — can transmit information of a sort from one extremity to the other. Thus we need not think that by assigning hypothetical functions to some fibers we have excluded them as subjects of further hypotheses. Several functions of protozoan bodies seem to demand the intervention of some means of information transfer more direct and precise than general cytoplasmic states or unchanneled diffusion of messenger substances. These include the careful programming of morphogenetic events, the coordinated contraction and relaxation of bodies or body parts, and, most conspicuously, ciliary and flagellar cooperation.

According to several recent accounts, membrane transmission of an excited state, in quite the conventional fashion, could account for changes of activity in the somatic ciliature of *Opalina* and several ciliates (Okajima, 1953, 1954a, 1954b; Naitoh, 1958 ; Párducz, 1958b; Jahn, 1960). According to other recent reports (e.g., the careful experiments of Sleigh, 1956, 1957) as well as a mass of older literature — little of it experimentally viable — the activity of specialized ciliature such as that in the adoral membranelles of *Stentor* requires a more subtle explanation. It is certain that no fibrillar system described for *Opalina, Paramecium,* and the tetrahymenids could, by its distribution, be held responsible for either normal metachrony or ciliary responses to stimuli over the general body surface; hence the membrane-transmission theory gains plausibility by default. Oral ciliature is something else again, since there are usually enough interconnecting elements to accommodate any theory, if one is not particular about details of experimental proof. The greatest obstacle to immediate progress is the almost prohibitive difficulty of determining by electron microscopy what ultrastructural constituents have and have not been altered by experimental manipulation, at least with the techniques devised to date.

A serious consideration of ciliary coordination would require scrutiny of metazoan systems and much more space than is available here — we can only acknowledge the problem.

Many of the fibrillar structures of protozoa are so placed as to make very tempting the idea that they could serve in the establishment and maintenance of spatial relationships, if one assumes that activities or states in the kinetosome could trigger activities or states in whatever is present at the opposite end of the fibril. In the complete absence of developmental studies of ultrastructure, such an idea remains tenuous; we have no way of distinguishing symptoms from causes.

Protozoan Relationships

The contribution of electron microscopy to protozoan taxonomy already is considerable. Among phytoflagellates that have so little microscopically visible structure at all, such properties as scale pattern and the number and morphology of flagella have justified revisions of generic and even familial and ordinal classification. The discovery of peculiar anterior organelles in several sporozoa clearly indicates the value of ultrastructural studies here. Of the larger and more complex protozoa, not enough species have been studied to suggest any modification of already abundant morphologic criteria. Among the euglenoids, trypanosomids, bodonids, gymnostomes, hymenostomes, heterotrichs, and entodiniomorphs, related genera have shown such similarities as one would expect.

As for detective work in the field of phylogenetic relationships among larger categories, it seems that, at the lower end of the scale, the biochemists will continue to have most of the fun.

Morphological trademarks are relatively few. Flagellar garnishment is certainly one of these, and the possession (but not the lack) of the ochromonad type of pantoneme flagellum seems to be a reliable symptom of kinship. In the phytomonads, the occurrence of mastigonemes is sporadic and inconsistent with any other suspected index of relationship. Studies of chloroplast and pyrenoid structure, as these become more numerous and precise, will undoubtedly yield additional exciting clues when combined with chemical analysis of photosynthetic pigments and products. The stigma–flagellum coaptation is another character that bears watching. It has been described in chrysomonads and brown-algal zoospores and in euglenids, but not yet in any phytomonad.

But the majority of phytoflagellates has not indulged, on a large scale, in morphologic experimentation, which means simply that the very different things they do are done differently at a molecular level still beyond our powers of direct visualization.

The euglenoids are quite another kettle of fish. In them are found for the first time fairly elaborate cytoplasmic constructions that are not in the common cellular repertoire. They have well-developed systems of tubular fibrils underlying the pellicle and (perhaps different kinds) composing the pharyngeal rodorgan in *Peranema*. The tubular fibril is probably not original with them. Spermatozoids of three land plants (*Sphagnum* — Manton, 1957a; *Pteridium* — Manton, 1959c; *Marchantia* — Heitz, 1959) contain orderly bands of tubular fibrils. Thus it would not be surprising to discover this type of fibril at least in the phytomonads. But apparently it remained to the euglenoids, among phytoflagellates, to discover just how useful it could be.

Several points of resemblance between the bodonids-trypanosomids and the euglenoids have been noted on p. 143. The most conspicuous is the system of pellicular fibrils connecting with the kinetosomes. In neither group have banded intracellular fibrils been found. Although the bodonid's single Golgi apparatus occupies a position near the kinetosomes (as in chrysomonads and phytomonads, but not in the euglenoids with their numerous dictyosomes), no rhizoplast-like fibrils have been reported as yet. It would be unwise to overemphasize this point, as flagellar rootlets in the phytomonads do not always appear banded, and the difference may be insignificant. At any rate, periodic fibers reappear with a vengeance in the metamonads.

A search for evidences of relationship between the bodonids and the trichomonads, supposed to be successive stem groups in the zooflagellate series, is not particularly rewarding at the moment. For one thing, the mysterious kinetoplast-mitochondrion of the bodonid-trypanosomid appears to be a unique structure. For another, the pellicular fibril system does not reappear in similar form anywhere among the higher zooflagellates yet seen.

With the trichomonads, there is a sudden flowering of zooflagellate architectural ingenuity. They have the striated parabasal fibril seen again in *Trichonympha* (the axial ribbons of *Pyrsonympha* and *Joenia* probably are the same thing put to a slightly different use); the fibrillar axostyle wall found in *Lophomonas* and *Holomastigotoides;* and the beautifully periodic costa, whose suspected counterparts occur in some other metamonads not yet studied. The occurrence of a centriole separate from the flagellum-producing ones has not been demonstrated by electron microscopy in *Trichomonas,* but is suggested in *Foaina.* Higher metamonads work with essentially the same construction units as do the trichomonads, but have conceived some astonishing ways of putting them together.

On the rhizopod side of flagellate relationships, the scarcity of evidence concerning flagellate stages makes speculation nearly impossible. Polyphyletic origin of various rhizopod and actinopod groups, as well as of slime molds, sporozoa, and fungi, is probable. The pellicular fibril system of *Tetramitus,* suggestively similar to that of euglenoids and bodonids, is noteworthy and needs study.

Not surprisingly, the study of ciliate ultrastructure does not yet offer any clues to ciliate origins. The simplest forms (if indeed there are any simple enough) have not been examined. Once again, their banded fibrils are conspicuously linked to kinetosomes. The fibrils look remarkably like the parabasal fibrils of some zooflagellates, but if there is any homology here it must be peculiarly devious, the ciliate kinetodesmal fibril having come unhooked from Golgi apparatus or nucleus and assuming instead a peripheral linear disposition. Perhaps a more direct homology is to be sought in the banded root fibrils of phytoflagellates. In all instances, the fibril is heteropolar and could have some significance in determining spatial relationships. Ciliates have carried to an impressive extreme the elaboration of tubular-fibrillar structures, with and without evident kinetosome attachments; these include orderly arrangements of pellicular fibrils reminiscent of the euglenoid-bodonid type.

In both the zooflagellates and the ciliates, the evolutionary trend appears to have been one toward increasing architectural complexity primarily involving fibrillar structures. It is at least interesting to point out that the straight, orderly, tubular fibril has not, to the author's knowledge, been described from any metazoan cell, except in cilia, kinetosomes, and centrioles. Metazoa have fibrillar structures — spindle fibers, muscle filaments, tonofilaments, neurofilaments, ciliary roots, and connective tissue components. Of these only the spindle fibers and ciliary apparatus are kinetosome-linked, and only the highly specialized striated muscle filaments are packed with the conspicuous precision characteristic of so many protozoan organelles. Metazoa have gone in for membrane elaborations and these by their very nature cannot assume the extraordinary geometric variety that is possible for fibers. No metazoan cell approaches in architectural complexity the design of the "simplest" ciliate or metamonad.

One gains the impression that the ancestors of zooflagellates and ciliates (whether or not these were the same, and whether or not the euglenoid type was involved) committed themselves (and their kinetosomes) to creative fibrillogenesis, and having done so shortly passed a point of no return. Fibrils could be made to answer many special requirements of an independent cell and did so with outstanding success. But there is no evidence from electron microscopy to suggest that this led to evolution beyond the protozoan level.

Perpetuation of an extremely complex and individualistic cell pattern must place severe demands on the cell's genetic mechanism. Perhaps the elaborate centriolar structures of the metamonads and the heterokaryote condition of ciliates are devices to assist in meeting these demands. It is a commonplace of metazoan morphogenesis that cells with a high degree of morphologic differentiation are not capable of continued division. But ciliates and metamonads *do* divide, temporarily suppressing and then reproducing their flamboyant individualities. It is conceivable that a single genetic system is incapable of meeting these demands and at the same time providing the delicate set of checks and balances necessary for intercellular cooperation in a metazoan organism. The author finds it impossible to visualize with Hadẑi and Hanson (see Hanson, 1958) the subdivision of a ciliate-like organism into cellular compartments as a step in metazoan evolution. Comparison of the protozoan with the total metazoan organism does not help, because no known systems of structures in metazoan tissues seem related, morphologically, to ciliate intracellular fiber systems. But certainly, the ultrastructure of lower metazoan ciliated epithelia and of the morphologically simple acoel turbellaria needs to be investigated.

The incompatibility of cellular complexity and multicellular differentiation is by no means a fresh idea. The only original notion here is that ultrastructure may give us a clue to the nature of a choice that was open to organisms at a protozoan level in evolution. Phytoflagellates, ameboflagellates, and the simplest protomonads possess all the fundamentals of metazoan (and protozoan and metaphytan) cell structure. Higher zooflagellates and ciliates went on to exploit to the full the potentialities of intracellular fibrillogenesis. Those organisms that ultimately were to spawn the metazoa retained the kinetosome/centriole as a versatile, polarized, morphogenetic center but did not enslave it, and their genetic systems, in the service of immediate morphologic elaboration.

This discussion has been teleologic, and deliberately so. One can afford to be teleologic only when one knows very little, or a very great deal, about the facts, and we know very little. If these notions are attacked with vigor by future workers who are able to marshal more facts to their purpose, then they will have served their function.

APPENDIX

Systematic list of protozoan genera subjected to electron-microscope study

Only those genera mentioned in this book are listed; references to them may be found by consulting the index. A few genera, cited only briefly or incidentally in the electron-microscope literature, are omitted.

The system of classification used here is taken from Grassé (1952, 1953) and (for ciliates) from Corliss (1961). It is modified only by the imposition of uniform endings as specified by Hall (1953). Since Hall lists no superorders, the ending for this category is adopted from Stenzel (1950).

Classification	*Genera studied*
Phylum PROTOZOA	
Subphylum Rhizoflagellata	
Superclass Flagellata	
Class Phytomonadea	*Chlamydomonas, Pedinomonas, Micromonas, Platymonas, Sphaerella*
Xanthomonadea	—
Chloromonadea	—
Euglenea	*Euglena, Astasia, Rhabdomonas, Phacus, Peranema, Entosiphon*
Cryptomonadea	*Cryptomonas, Chilomonas, Hemiselmis*
Dinoflagellatea	*Gymnodinium, Amphidinium, Oxyrrhis, Gyrodinium, Polykrikos, Ceratium,* Zooxanthella

Class Ebriidea	—
Silicoflagellatea	—
Coccolithophorea	Several genera (scales and coccoliths only)
Chrysomonadea	*Chromulina, Monochrysis, Mallomonas, Hydrurus, Ochromonas, Poteriochromonas, Paraphysomonas, Synura, Chrysosphaerella, Stokesiella, Prymnesium, Chrysochromulina*
Class Zooflagellatea	
Superorder Protomonadica	
Order Choanoflagellida	*Codonosiga*
Biocoecida	*Bicoeca*
Trypanosomatida	*Leptomonas, Crithidia, Leishmania, Trypanosoma*
Bodonida	*Bodo, Cryptobia*
Proteromonadida	—
Superorder Metamonadica	
Order Trichomonadida	*Trichomonas, Tritrichomonas, Foaina*
Pyrsonymphida	*Pyrsonympha*
Oxymonadida	—
Retortamonadida	—
Joeniida	*Joenia*
Lophomonadida	*Lophomonas*
Trichonymphida	*Trichonympha, Pseudotrichonympha, Barbulanympha*
Spirotrichonymphida	*Holomastigotoides*
Superorder Opalinica	*Opalina*
Superclass Rhizopoda	
Class Lobosea	*Tetramitus, Amoeba, Pelomyxa, Hyalodiscus, Hartmannella, Acanthomoeba, Entamoeba, Endamoeba, Gromia*

Subphylum Actinopoda
  Class Acantharea   —
    Radiolarea   —
    Heliozoea   *Actinophrys,*
                                Actinosphaerium,
                                Heterophrys

Rhizopods? or Fungi?
  Order Acrasida   *Dictyostelium,*
                                Polysphondylium
    Mycetozoida   *Physarum, Didymium*
    Plasmodiophorida   —

Subphylum Sporozoa
  Class Gregarinomorphea   *Apolocystis, Gregarina,*
                                Stylocephalus, Lophocephalus,
                                Beloides
    Coccidiomorphea   *Plasmodium, Lankesterella*
    Sarcosporidea   *Sarcocystis*
    Unclassified   *Babesia, Toxoplasma,*
                                Anaplasma

Subphylum Cnidospora
  Order Myxosporida   —
    Microsporida   *Nosema, Plistophora*

Subphylum Ciliophora
  Class Ciliatea
    Subclass Holotrichia
      Order Gymnostomatida   *Coleps, Prorodon, Dysteria,*
                                  Chlamydodon, Nassula,
                                24 gymnostomes
                                (trichocysts only)
      Trichostomatida   *Isotricha*
      Chonotrichida   *Chilodochona*
      Suctorida   *Tokophrya, Ephelota,*
                                  Discophrya
      Apostomatida   —
      Astomatida   *Metaradiophrya*
      Hymenostomatida   *Ophryoglena, Colpidium,*
                                  Glaucoma, Tetrahymena,
                                Paramecium, Frontonia

Thigmotrichida   —

Peritrichida   *Ophrydium, Carchesium, Vorticella, Zoothamnium, Campanella, Epistylis, Opercularia, Trichodinopsis*

Subclass Spirotrichia

Order Heterotrichida   *Stentor, Blepharisma, Spirostomum, Nyctotherus, Condylostoma*

Oligotrichida   —

Tintinnida   —

Entodiniomorphida   Several ophryoscolecid genera

Odontostomatida   —

Hypotrichida   *Euplotes, Stylonychia*

REFERENCES

Afzelius, B. (1959) Electron microscopy of the sperm tail. Results obtained with a new fixative. *J. biophys. biochem. Cytol.,* **5,** 269–278.

—— (1961) The fine structure of the cilia from ctenophore swimming-plates. *J. biophys. biochem. Cytol.,* **9,** 383–394.

Allen, R. D. (1960) The consistency of ameba cytoplasm and its bearing on the mechanism of ameboid movement. II. The effects of centrifugal acceleration observed in the centrifuge microscope. *J. biophys. biochem. Cytol.,* **8,** 379–398.

—— (1961) Ameboid movement. *In* Brachet, J. and A. E. Mirsky (eds.), *The Cell,* **2,** 135–216. Academic Press, New York.

Allen, R. D., Cooledge, J. W., and P. J. Hall (1960) Streaming in cytoplasm dissociated from the giant amoeba, *Chaos chaos. Nature, Lond.,* **187,** 896–899.

Allen, R. D. and J. D. Roslansky (1959) The consistency of ameba cytoplasm and its bearing on the mechanism of ameboid movement. I. An analysis of endoplasmic velocity profiles of *Chaos chaos* (L). *J. biophys. biochem. Cytol.,* **6,** 437–446.

Anderson, E. (1955) The electron microscopy of *Trichomonas muris. J. Protozool.,* **2,** 114–124.

Anderson, E. and H. W. Beams (1959) The cytology of *Tritrichomonas* as revealed by the electron microscope. *J. Morphol.,* **104,** 205–235.

—— (1960) The fine structure of the heliozoan, *Actinosphaerium nucleofilum. J. Protozool.,* **7,** 190–199.

—— (1961) The ultrastructure of *Tritrichomonas* with special reference to the blepharoplast complex. *J. Protozool.,* **8,** 71–75.

Anderson, E., Saxe, L. H., and H. W. Beams (1956) Electron microscope observations of *Trypanosoma equiperdum. J. Parasit.,* **42,** 11–16.

André, J. (1959) Étude au microscope électronique de l'évolution du chondriome pendant la spermatogénèse du scorpion *Euscorpius flavicaudis. J. Ultrastruct. Res.,* **2,** 288–308.

—— (1961) Sur quelques détails nouvellement connus de l'ultrastructure des organites vibratiles. *J. Ultrastruct. Res.,* **5,** 86–108.

Asmund, B. (1955) Electron microscope observations on *Mallomonas caudata* and some remarks on its occurrence in four Danish ponds. *Bot. Tidsskr.,* **52,** 163–168.

—— (1959) Electron microscope observations on *Mallomonas* species. III. *Dansk bot. Ark.,* **18,** 3.

Astbury, W. T., Beighton, E., and C. Weibull (1955) The structure of bacterial flagella. *Symp. Soc. exp. Biol.,* **9,** 282–305.

BAIRATI, A. and F. E. LEHMANN (1952) Über die submikroskopische Struktur der Kernmembran bei *Amoeba proteus. Experientia,* **8,** 60–61.

—— (1953) Structural and chemical properties of the plasmalemma of *Amoeba proteus. Exp. Cell Res.,* **5,** 220–233.

—— (1954) Partial disintegration of cytoplasmic structures of *Amoeba proteus* after fixation with osmium tetroxide. *Experientia,* **10,** 173–178.

—— (1956) Structural and chemical properties of the contractile vacuole of *Amoeba proteus. Protoplasma,* **45,** 525–539.

BALL, G. H. (1960) Some considerations regarding the Sporozoa. *J. Protozool.,* **7,** 1–6.

BARER, R., JOSEPH, S., and G. A. MEEK (1959) The origin of the nuclear membrane. *Exp. Cell Res.,* **18,** 179–182.

BARKER, D. C. and K. DEUTSCH (1958) The chromatoid body of *Entamoeba invadens. Exp. Cell Res.,* **15,** 604–610.

BEALE, G. H. and A. JURAND (1960) Structure of the mate-killer (mu) particles in *Paramecium aurelia,* stock 540. *J. gen. Microbiol.,* **23,** 243–252.

BEAMS, H. W., KING, R. L., TAHMISIAN, T. N., and R. DEVINE (1960) Electron microscope studies on *Lophomonas striata* with special reference to the nature and position of the striations. *J. Protozool.,* **7,** 91–101.

BEAMS, H. W., TAHMISIAN, T. N., ANDERSON, E., and W. Wright (1961) Studies on the fine structure of *Lophomonas blattarum* with special reference to the so-called parabasal apparatus. *J. Ultrastruct. Res.,* **5,** 166–183.

BEAMS, H. W., TAHMISIAN, T. N., DEVINE, R. L., and E. ANDERSON (1957) Ultrastructure of the nuclear membrane of a gregarine parasitic in grasshoppers. *Exp. Cell Res.,* **13,** 200–204.

—— (1959a) Studies on the fine structure of a gregarine parasitic in the gut of the grasshopper, *Melanoplus differentialis. J. Protozool.,* **6,** 136–146.

—— (1959b) The fine structure of the nuclear envelope of *Endamoeba blattae. Exp. Cell Res.,* **18,** 366–369.

BENNETT, H. S. (1956) The concepts of membrane flow and membrane vesiculation as mechanisms for active transport and ion pumping. *J. biophys. biochem. Cytol.,* **2** (Suppl.), 99–103.

BERGERON, J. A. (1959) The bacterial chromatophore. *Brookhaven Symp. Biol.,* **11,** 118–131.

BERNHARD, W. (1959) Ultrastructural aspects of nucleocytoplasmic relationship. *Exp. Cell Res.,* Suppl. **6,** 17–50.

BERNHARD, W. and E. DE HARVEN (1960) L'ultrastructure du centriole et d'autres éléments de l'appareil achromatique. *Verhandl. IV intern. Kong. Elektronenmikroskopie, Berlin,* 1958, **2,** 217–227.

BEYERSDORFER, K. and J. DRAGESCO (1952a) Étude comparative des trichocystes de sept éspèces de paramécies. *Congr. Microscopie Électronique, Édit. Revue d'Optique,* Paris, pp. 661–671.

—— (1952b) Microscopie électronique des trichocystes de *Frontonia. Congr. Microscopie Électronique, Édit. Revue d'Optique,* Paris, pp. 655–660.

BISHOP, D. W. and H. HOFFMANN-BERLING (1959) Extracted mammalian sperm models. I. Preparation and reactivation with adenosine triphosphate. *J. cell. comp. Physiol.*, 53, 445–466.

BLANC-BRUDE, R., DRAGESCO, J., and J. P. HERMET (1951) Microscopie électronique des tentacules de l'acinétien *Discophrya piriformis* Guilcher. *Bull. Microscop. appl.*, 1, 29–30.

BLANCKART, S. (1957) Die Oberflächenstrukturen von *Paramecium* spec. und *Opalina ranarum. Z. wiss. Mikroskop.*, 63, 276–287.

BORYSKO, E. and J. ROSLANSKY (1959) Methods for correlated optical and electron microscopic studies of amoebae, *Ann N.Y. Acad. Sci.*, 78, 432–447.

BRACHET, J. (1957) *Biochemical Cytology*. Academic Press, New York. 516 pp.

―――― (1958) New observations on biochemical interactions between nucleus and cytoplasm in *Amoeba* and *Acetabularia. Exp. Cell Res.*, Suppl. 6, 78–96.

―――― (1959) Cytoplasmic dependence in amoebae. *Ann. N.Y. Acad. Sci.*, 78, 688–695.

BRACHET, J. and A. E. MIRSKY (eds.) (1959–1961) *The Cell*, Vols. I–V. Academic Press, New York.

BRADFIELD, J. R. G. (1955) Fiber patterns in animal flagella and cilia. *Symp. Soc. exp. Biol.*, 9, 306–334.

BRANDT, P. W. (1958) A study of the mechanism of pinocytosis. *Exp. Cell Res.*, 15, 300–313.

BRANDT, P. W. and G. D. PAPPAS (1959) Mitochondria. II. The nuclear-mitochondrial relationship in *Pelomyxa carolinensis* Wilson (*Chaos chaos* L.). *J. biophys. biochem. Cytol.*, 6, 91–96.

―――― (1960) An electron microscopic study of pinocytosis in ameba. I. The surface attachment phase. *J. biophys. biochem. Cytol.*, 8, 675–687.

BRETSCHNEIDER, L. H. (1950) Elektronenmikroskopische Untersuchung einiger Ziliaten. *Mikroskopie*, Wien 5, 257–259.

―――― (1959) Die submikroskopische Struktur der Pellikula von *Epidinium ecaudatum*, (Ophryoscolecidae). *Proc. Acad. Sci. Amst.* (Ser. C) 62, 542–555.

―――― (1960) Elektronenmikroskopische Untersuchung des Peristom-apparates einiger Ophryoscoleciden. *Proc. Acad. Sci. Amst.* (Ser. C) 63, 291–317.

BRODY, M. and A. E. VATTER (1959) Observations on cellular structures of *Porphyridium cruentum. J. biophys. biochem. Cytol.*, 5, 289–294.

BROWN, H. P. (1945) On the structure and mechanics of the protozoan flagellum. *Ohio J. Sci.*, 45, 247–301.

BROWN, H. P. and A. COX (1954) An electron microscope study of protozoan flagella. *Amer. Midl. Nat.*, 52, 106–117.

BÜTSCHLI, O. (1887–1889) *Protozoa*. Abt. III. *Infusoria und System der Radiolaria. In* BRONN, H. G. (ed.), *Klassen und Ordnung des Thier-reichs.* 1, 1098–2035. C. F. Winter, Leipzig.

CALVIN, M. (1959) From microstructure to macrostructure and function in the photochemical apparatus. *Brookhaven Symp. Biol.*, **11**, 160–179.

CHAKRABORTY, J. and N. N. DAS GUPTA (1960) Ultrastructure of the pellicle and the nucleus of *Leishmania donovani*. *Verhandl. IV intern. Kong. Elektronenmikroskopie, Berlin,* 1958, **2**, 510–515.

CHANG, P. C. H. (1956) The ultrastructure of *Leishmania donovani*. *J. Parasitol.,* **42**, 126–136.

CHAPMAN-ANDRESEN, C. and J. R. NILSSON (1960) Electron micrographs of pinocytosis channels in *Amoeba proteus*. *Exp. Cell Res.,* **19**, 631–633.

CHATTON, E. (1952) Classe des Dinoflagellés ou Péridiniens. *In* GRASSÉ, P.P. (ed.) *Traité de Zoologie,* T.1, F. Vol. 1, 309–406, Masson et Cie., Paris.

CHATTON, E. (1953) Classe des Lobosa. Ordre des amoebiens nus ou Amoebaea. *In* GRASSÉ, P. P. (ed.), *Traité de Zoologie,* T. 1, F.,Vol. 2, 5–91, Masson et Cie., Paris.

CHATTON, E. and A. LWOFF(1935) Les ciliés apostomes. I. Aperçu historique et général; étude monographique des genres et des éspèces. *Arch. Zool. exptl. gén.,* **77**, 1–453.

CHEN, Y. T. (1950a) Investigations into the biology of *Peranema trichophorum* (Euglenineae). *Quart. J. micr. Sci.,* **91**, 279–308.

—— (1950b) The flagellar structure of some protista. *Proc. Conf. Electron Microscopy,* Delft, 1949, pp. 156–158.

CHILD, F. M. (1959) The characterization of the cilia of *Tetrahymena pyriformis*. *Exp. Cell Res.,* **18**, 258–267.

CIGADA, M. L. and C. CANTONE (1960) Growth of the gametes of *Amoeba spumosa* Grübner. *Nature, Lond.,* **188**, 1046–1047.

CLARK, T. B. (1959) Comparative morphology of four genera of the Trypanosomatidae. *J. Protozool.,* **6**, 227–232.

CLARK, T. B. and F. G. WALLACE (1960) A comparative study of kinetoplast ultrastructure in the Trypanosomatidae. *J. Protozool.,* **7**, 115–124.

CLEVELAND, L. R. (1957) Types and life cycles of centrioles of flagellates. *J. Protozool.,* **4**, 230–241.

—— (1960) The centrioles of *Trichonympha* from termites and their functions in reproduction. *J. Protozool.,* **7**, 326–341.

COHEN, A. I. (1957) Electron microscopic observations of *Amoeba proteus* in growth and inanition. *J. biophys. biochem. Cytol.,* **3**, 859–866.

COPELAND, H. F. (1956) *The Classification of Lower Organisms.* Pacific Books, Palo Alto. 302 pp.

CORLISS, J. O. (1955) The opalinid infusorians: flagellates or ciliates? *J. Protozool.,* **2**, 107–114.

—— (1956) On the evolution and systematics of ciliated protozoa. *Syst. Zool.,* **5**, 68–91, 121–140.

—— (1958) The phylogenetic significance of the genus *Pseudomicrothorax* in the evolution of holotrichous ciliates. *Acta Biol. Acad. Sci. Hung.,* **8**, 367–388.

—— (1959a) An illustrated key to the higher groups of the ciliated protozoa with definition of terms. *J. Protozool.,* **6**, 265–281.

CORLISS, J. O. (1959b) Comments on the systematics and phylogeny of the Protozoa. *Syst. Zool.,* **8,** 169–190.

———— (1961) *The Ciliated Protozoa.* Pergamon Press, London. 310 pp.

DAS GUPTA, N. N., BATTACHARYA, D. L., and P. C. SEN GUPTA (1951) Electron and photomicrographic studies of the flagellate form of *Leishmania donovani. Nature, Lond.,* **167,** 1063.

DAS GUPTA, N. N., GUHA, A., and M. L. DE (1954) Observations on the structure of *Leishmania donovani* — the Kala-azar parasite. *Exp. Cell Res.,* **6,** 353–360.

DEDEKEN-GRENSON, M. (1960) Le méchanisme de la perte héréditaire d'une fonction biochimique: la synthèse des chlorophylles. *Arch. Biol., Paris,* **71,** 269–342.

DEFLANDRE, G. (1953) Ordres des Testacealobosa, des Testaceafilosa, des Thalamia ou Thécamoebiens. *In* GRASSÉ, P. P. (ed.), *Traité de Zoologie,* T. 1, F., **2,** 97–148, Masson et Cie, Paris.

DEFLANDRE, G. and C. FERT (1953) Étude des coccolithophoridés des vases actuelles au microscope électronique. *C. R. Acad. Sci., Paris,* **236,** 328–330.

DEROBERTIS, E. (1956) Electron microscope observations on the submicroscopic organization of the retinal rods. *J. biophys. biochem. Cytol.,* **2,** 319–330.

DEUTSCH, K. and M. M. SWANN (1959) An electron microscope study of a small free-living amoeba (*Hartmannella astronyxis*). *Quart. J. micr. Sci.,* **100,** 13–15.

DEUTSCH, K. and V. ZAMAN (1959) An electron microscope study of *Entamoeba invadens* Rodhain 1934. *Exp. Cell Res.,* **17,** 310–319.

DINICHERT, P., GUYÉNOT, E., and M. ZALOKAR (1947) Observations cytologiques avec le microscope électronique. *Rev. suisse Zool.,* **54,** 283–290.

DIPPELL, R. V. (1958) The fine structure of kappa in killer stock 51 of *Paramecium aurelia. J. biophys. biochem. Cytol.,* **4,** 125–128.

DOUGHERTY, E. C. (1955) Comparative evolution and the origin of sexuality. *Syst. Zool.,* **4,** 145–169, 190.

———— (1957a) Neologisms needed for structures of primitive organisms. 1. Types of nuclei. *J. Protozool.,* 4 (Suppl.), 14.

———— (1957b) Neologisms needed for structures of primitive organisms. 2. Vibratile organelles. *J. Protozool.,* 4 (Suppl.), 14.

DOUGHERTY, E. C. and M. B. ALLEN (1958) The words "protist" and "Protista". *Experientia,* **14,** 78–81.

———— (1959) Speculations on the position of the cryptomonads in protistan phylogeny. *Proc. XVth intern. Congr. Zool., London,* 1958, pp. 184–186.

———— (1960) Is pigmentation a clue to protistan phylogeny? In ALLEN, M.B. (ed.), *Comparative Biochemistry of Photoreactive Systems,* pp. 129–144. Academic Press, New York.

DRAGESCO, J. (1951) Sur la structure des trichocystes du flagellé crytomonadine *Chilomonas paramecium. Bull. Microscop. appl.,* **1,** 172–175.

DRAGESCO, J. (1952a) Electron microscopy of the trichocysts of the holotrichous ciliates *Nassula elegans* and *Disematostoma bütschlii. Proc. Soc. Protozool.,* **3,** 15.

—— (1952b) Le flagellé *Oxyrrhis marina:* cytologie, trichocystes, position systématique. *Bull Microscop. appl.,* **2,** 148–157.

—— (1952c) Sur la structure des trichocystes toxiques des infusoires holotriches gymnostomes. *Bull. Microscop. appl.,* **2,** 92–98.

—— (1952d) The mucoid trichocysts of flagellates and ciliates. *Proc. Soc. Protozool.,* **3,** 15.

DUNCAN, D., EADES, J., JULIAN, S. R., and D. O. MICKS (1960) Electron microscope observations on malarial oocysts (*Plasmodium cathemerium*). *J. Protozool.,* **7,** 18–26.

DUNCAN, D., STREET, J., JULIAN, S. R., and D. O. MICKS (1959) Electron microscope observations on the gametocytes of a malarial parasite (*Plasmodium cathemerium*). *Texas Rep. Biol. Med.,* **17,** 314–322.

EAKIN, R. M. and J. A. WESTFALL (1959) Fine structure of the retina in the reptilian third eye. *J. biophys. biochem. Cytol.,* **6,** 133–134.

EHRET, C. F. (1958) Information content and biotopology of the cell in terms of cell organelles. *In* YOCKEY, H. P., PLATZMAN, R. L., and H. QUASTLER (eds.), *Symposium on Information Theory in Biology,* pp. 218–229. Pergamon Press, Oxford.

—— (1960) Organelle systems and biological organization. *Science,* **132,** 115–123.

EHRET, C. F. and E. L. POWERS (1955) Marconuclear and nucleolar development in *Paramecium bursaria. Exp. Cell Res.,* **9,** 241–257.

—— (1957) The organization of gullet organelles in *Paramecium bursaria. J. Protozool.,* **4,** 55–59.

—— (1959) The cell surface of *Paramecium. Int. Rev. Cytol.,* **8,** 97–133.

ELLIOTT, A. M. and J. W. TREMOR (1958) The fine structure of the pellicle in the contact area of conjugating *Tetrahymena pyriformis. J. biophys. biochem. Cytol.,* **4,** 839–840.

ESPANA, C., ESPANA, E. M., and D. GONZALEZ (1959) *Anaplasma marginale.* I. Studies with phase contrast and electron microscopy. *Amer. J. vet. Res.,* **20,** 795–805.

FAURÉ-FREMIET, E. (1950) Morphologie comparée et systématique des Ciliés. *Bull. Soc. zool. Fr.,* **75,** 109–122.

—— (1953) Morphology of protozoa. *Annu. Rev. Microbiol.,* **7,** 1–18.

—— (1957) Finer morphology of microorganisms. *Annu. Rev. Microbiol.,* **11,** 1–6.

FAURÉ-FREMIET, E. and C. ROUILLER (1958a) Myonèmes et cinétodesmes chez les ciliés du genre *Stentor. Bull. Microscop. appl.,* **8,** 117–119.

—— (1958b) Réseau canaliculaire dans les myonèmes endoplasmiques de quelques ciliés. *C. R. Acad. Sci., Paris,* **246,** 2039–2042.

—— (1959) Le cortex de la vacuole contractile et son ultrastructure chez les ciliés. *J. Protozool.,* **6,** 29–37.

FAURÉ-FREMIET, E., ROUILLER, C., and M. GAUCHERY (1956a) L'appareil squelettique et myoide des urcéolaires: étude au microscope électronique. *Bull. Soc. zool. Fr.,* **81,** 77–84.

——— (1956b) Les structures myoides chez les ciliés. Étude au microscope électronique. *Arch. Anat. microscop.,* **45,** 139–161.

——— (1956c) Structure et origine du pedoncule chez *Chilodochona. J. Protozool.,* **3,** 188–193.

——— (1957) La réorganisation macronucléaire chez les *Euplotes. Exp. Cell Res.,* **12,** 135–144.

FAWCETT, D. W. (1958) The structure of the mammalian spermatozoon. *Int. Rev. Cytol.,* **7,** 195–235.

——— (1961) Cilia and flagella. *In* BRACHET, J. and A. E. MIRSKY (eds.), *The Cell,* **2,** 217–298. Academic Press, New York.

FAWCETT, D. W. and K. R. PORTER (1954) A study of the fine structure of ciliated epithelia. *J. Morphol.,* **94,** 221–282.

FINCK, H. and H. HOLTZER (1961) Attempts to detect myosin and actin in cilia and flagella. *Exp. Cell Res.,* **23,** 251–257.

FINLEY, H. E. (1951) Electron microscopy of thin-sectioned *Spirostomum. Science,* **113,** 362–363.

——— (1955) Electron microscopical observations on *Spiristomum ambiguum. Ann. N.Y. Acad. Sci.,* **62,** 229–246.

FINLEY, H. E. and C. A. BROWN (1960) Ultrastructure in a ciliated protozoan. *Norelco Reporter,* **7,** 24–28, 66.

FJERDINGSTAD, E. J. (1961a) The ultrastructure of choanocyte collars in *Spongilla lacustris* (L). *Z. Zellforsch.,* **53,** 645–657.

——— (1961b) Ultrastructure of the collar of the choanoflagellate *Codonosiga botrytis* (Ehrenb.), *Z. Zellforsch,* **54,** 499–510.

FOOTE, L. E., GEER, J. C., and Y. E. STICH (1958) Electron microscopy of the *Anaplasma* body: ultrathin sections of bovine erythrocytes. *Science,* **128,** 147–148.

FOSTER, E., BAYLOR, M. B., MEINKOTH, N. A., and G. L. CLARK (1947) An electron microscope study of protozoan flagella. *Biol. Bull. Wood's Hole,* **93,** 114–121.

FOTT, B. (1955) Scales of *Mallomonas* observed in the electron microscope. *Preslia,* **27,** 280–282.

FOTT, B. and J. LUDVIK (1956a) Elektronenoptische Untersuchung der Kieselstrukturen bei *Chrysosphaerella* (Chrysomonadineae). *Preslia,* **28,** 276–278.

——— (1956b) Über den submikroskopischen Bau des Panzers von *Ceratium hirundinella. Preslia,* **28,** 278–280.

FRITSCH, F. E. (1945) *The Structure and Reproduction of the Algae.* Vol. II, Cambridge Univ. Press, London, 939 pp.

FULTON, J. D. and T. H. FLEWETT (1956) The relation of *Plasmodium berghei* and *Plasmodium knowlesi* to their respective red-cell hosts. *Trans. R. Soc. trop. Med. Hyg.,* **50,** 150–156.

Gaffron, H. (1960) The origin of life. *In* Tax, S. (ed.), *Evolution after Darwin*, **1**, 39–84. Univ. Chicago Press, Chicago.

Gall, J. G. (1959) Macronuclear duplication in the ciliated protozoan *Euplotes*. *J. biophys. biochem. Cytol.*, **5**, 295–308.

—— (1961) Centriole replication. A study of spermatogenesis in the snail *Viviparus*. *J. biophys. biochem. Cytol.*, **10**, 163–194.

Garnham, P. C. C., Bird, R. G., and J. R. Baker (1960) Electron microscope studies of motile stages of malaria parasites. I. The fine structure of the sporozoites of *Haemamoeba* (= *Plasmodium*) *gallinacea*. *Trans. R. Soc. trop. Med. Hyg.*, **54**, 274–278.

Garnham, P. C. C., Bird, R. G., Baker, J. R., and R. S. Bray (1961) Electron microscope studies of motile stages of malaria parasites. II The fine structure of the sporozoite of *Laverania* (= *Plasmodium*) *falcipara*. *Trans. R. Soc. trop. Med. Hyg.*, **55**, 98–102.

Gatenby, J. B., Dalton, A. J., and M. D. Felix (1955) The contractile vacuole of Parazoa and Protozoa and the Golgi apparatus. *Nature, Lond.* **176**, 301–306.

Gelei, J. v. (1936) Das erregungsleitende System der Ciliaten. *C. R. XII, Congr. intern. Zool.*, Lisbonne, pp. 174–206.

Gezelius, K. (1961) Further studies in the ultrastructure of Acrasiae. *Exp. Cell Res.*, **23**, 300–310.

Gezelius, K. and B. G. Rånby (1957) Morphology and fine structure of the slime mold *Dictyostelium discoideum*. *Exp. Cell Res.*, **12**, 265–284.

Gibbons, I. R. (1961) Structural asymmetry in cilia and flagella. *Nature, Lond.* **190**, 1128–1129.

Gibbons, I. R. and A. V. Grimstone (1960) On flagellar structure in certain flagellates. *J. biophys. biochem. Cytol.*, **7**, 697–716.

Gibbs, S. P. (1960) The fine structure of *Euglena gracilis* with special reference to the chloroplasts and pyrenoids. *J. Ultrastruct. Res.*, **4**, 127–148.

Gibbs, S. P., Lewin, R. A., and D. E. Philpott (1958) The fine structure of the flagellar apparatus of *Chlamydomonas moewusii*. *Exp. Cell Res.*, **15**, 619–622.

Grassé, P. P. (ed.) (1952) *Traité de Zoologie*. Vol. I, fasc. 1. *Phylogénie. Protozoaires: Généralités, Flagellés*. Masson et Cie., Paris. 1071 pp.

—— (ed.) (1953) Traité de Zoologie. Vol. I, fasc. 2. *Protozoaires: Rhizopodes, Actinipodes, Sporozoaires, Cnidosporidies*. Masson et Cie, Paris. 1160 pp.

—— (1956a) L'appareil parabasal et l'appareil de Golgi sont un même organite. Leur ultrastructure, leurs modes de sécrétion. *C. R. Acad. Sci., Paris*, **242**, 858–861.

—— (1956b) L'ultrastructure de *Pyrsonympha vertens* (Zooflagellata Pyrsonymphina): les flagelles et leur coaptation avec le corps, l'axostyle contractile, le paraxostyle, le cytoplasme. *Arch. Biol., Paris*, **67**, 595–611.

—— (1957a) L'appareil de Golgi des protozoaires et son ultrastructure comparée à celle des métazoaires. *Proc. Stockholm Conf. Electron Microscopy* 1956, pp. 143–145.

GRASSÉ, P. P. (1957b) Structure cellulaire et microscopie électronique. *Ann. Rech. med.*, 33, 617–632.

—— (1957c) Ultrastructure, polarité et reproduction de l'appareil de Golgi. *C. R. Acad. Sci., Paris*, 243, 1278–1281.

GRASSÉ, P. P. and N. CARASSO (1957) Ultra-structure of the Golgi apparatus in protozoa and metazoa (somatic and germinal cells). *Nature, Lond.*, 179, 31–33.

GRASSÉ, P. P. and G. DEFLANDRE (1952) Ordre des Bicoecidae. *In* GRASSÉ, P. P., *Traité de Zoologie*, T. 1, f., 1, 599–601. Masson et Cie., Paris.

GRASSÉ, P. P. and J. DRAGESCO (1957) L'ultrastructure du chromosome des péridiniens et ses conséquences génétiques. *C. R. Acad. Sci., Paris*, 245, 2447–2452.

GRASSÉ, P. P. and A. HOLLANDE (1941) Vacuoles pulsatiles et appareil de Golgi dans l'évolution de la cellule. *Arch. Zool. exptl. gén.*, 82, 301–319.

GRASSÉ, P. P. and H. MUGARD (1961) Les organites mucifères et la formation du kyste chez *Ophryoglena mucifera* (infusoire holotriche). *C. R. Acad. Sci., Paris.*, 253, 31–34.

GRASSÉ, P. P. and J. THÉODORIDÈS (1957) L'ultrastructure de la membrane nucléaire des grégarines. *C. R. Acad. Sci., Paris*, 245, 1985–1986.

—— (1958) La présence de l'ergastoplasme chez les protozoaires (cas des grégarines). *C. R. Acad. Sci., Paris*, 246, 1352–1353.

—— (1959) Recherches sur l'ultrastructure de quelques grégarines. *Ann. Sci. nat. Zool.*, 12, 237–252.

GRAY, J. (1955) The movement of sea-urchin spermatozoa. *J. exp. Biol.*, 32, 775–801

GREEN, D. E. (1959) Mitochondrial structure and function. *In* HAYASHI, T. (ed.), *Subcellular Particles*, pp. 84–103. Ronald Press, New York.

GREEN, D. E. and Y. HATEFI (1961) The mitochondrion and biochemical machines. *Science*, 133, 13–19.

GREENBERG, M. J. (1959) Ancestors, embryos and symmetry. *Syst. Zool.*, 8, 212–221.

GREENWOOD, A. D. (1959) Observations on the structure of the zoospores of *Vaucheria*. II. *J. exp. Bot.*, 10, 55–68.

GREIDER, M. H., KOSTIR, W. J., and W. J. FRAJOLA (1958) Electron microscopy of *Amoeba proteus*. *J. Protozool.*, 5, 139–146.

GRELL, K. G. V. (1956) *Protozoologie*. Springer Verlag, Berlin. 284 pp.

GRELL, K. G. V. and K. E. WOHLFARTH-BOTTERMANN (1957) Licht und electronenmikroskopische Untersuchungen an den Dinoflagellaten *Amphidinium elegans n.* sp. *Z. Zellforsch.*, 47, 7–17.

GRIFFIN, J. L. (1960) The isolation, characterization, and identification of the crystalline inclusions of the large free-living amebae. *J. biophys. biochem. Cytol.*, 7, 227–234.

GRIFFIN, J. L. and R. D. ALLEN (1960) The movement of particles attached to the surface of amebae in relation to current theories of ameboid movement. *Exp. Cell Res.*, 20, 619–622.

GRIMSTONE, A. V. (1959a) Aspects of cytoplasmic organization in *Trichonympha*. *Proc. XVth intern. Congr. Zool., London*, 1958, pp. 480–482.

—— (1959b) Cytology, homology and phylogeny — a note on "organic design". *Amer. Nat.*, **93**, 273–282.

—— (1959c) Cytoplasmic membranes and the nuclear membrane in the flagellate *Trichonympha*. *J. biophys. biochem. Cytol.*, **6**, 369–378.

—— (1961) Fine structure and morphogenesis in protozoa. *Biol. Rev.*, **36**, 97–150.

GROUPÉ, V. (1947) Surface striations of *Euglena gracilis* revealed by electron microscopy. *Proc. Soc. exp. Biol., N.Y.*, **64**, 401–403.

GRUNBAUM, B. W., MØLLER, K. M., and R. S. THOMAS (1959) Cytoplasmic crystals of the amoebae, *Amoeba proteus* and *Chaos chaos*. *Exp. Cell Res.*, **18**, 385–389.

GUSTAFSON, P. V., AGAR, H. D., and D. I. CROMER (1954) An electron microscope study of *Toxoplasma*. *Amer. J. trop. Med. Hyg.*, **3**, 1008–1021.

GUTTES, E. and S. GUTTES (1960) Pinocytosis in the myxomycete *Physarum polycephalum*. *Exp. Cell Res.*, **20**, 239–241.

HAGENAU, F. (1958) The ergastoplasm: its history, ultrastructure and biochemistry. *Int. Rev. Cytol.*, **7**, 425–483.

HALL, R. P. (1953) *Protozoology*. Prentice-Hall, New York, 682 pp.

HALLER, G. DE (1959) Structure submicroscopique d'*Euglena viridis*. *Arch. Sci.*, **12**, 307–348.

—— (1960) Contribution à la cytologie d'*Euglena viridis*. *Verhandl. IV intern. Kong. Elektronenmikroskopie, Berlin*, 1958, **2**, 517–520.

HAMILTON, L. D. and M. E. GETTNER (1958) Fine structure of kappa in *Paramecium aurelia*. *J. biophys. biochem. Cytol.*, **4**, 122–124.

HANSON, E. D. (1958) On the origin of the Eumetazoa. *Syst. Zool.*, **7**, 16–47.

HARRIS, K. and D. E. BRADLEY (1958) Some unusual Chrysophyceae studied in the electron microscope. *J. gen. Microbiol.*, **18**, 71–83.

—— (1960) A taxonomic study of *Mallomonas*. *J. gen Microbiol.*, **22**, 750–777.

HARRIS, P. (1961) Electron microscope study of mitosis in sea urchin blastomeres. *J. biophys. biochem. Cytol.*, **11**, 419–432.

HARRIS, P. and T. W. JAMES (1952) Electron microscope study of the nuclear membrane of *Amoeba proteus* in thin section. *Experientia*, **8**, 384–385.

HARRIS, P. and D. MAZIA (1959) The use of mercuric bromphenol blue as a stain for electron microscopy. *J. biophys. biochem. Cytol.*, **5**, 343–344.

HARVEN, E. DE and W. BERNHARD (1956) Étude au microscope électronique de l'ultrastructure du centriole chez les vertebrés. *Z. Zellforsch.*, **45**, 378–398.

HEDLEY, R. H. and W. S. BERTAUD (1962) Electron-microscopic observations of *Gromia oviformia* (Sarcodina). *J. Protozool.* **9**, 79–87.

HEITZ, E. (1959) Elektronenmikroskopische Untersuchungen über zwei auffallende Strukturen an der Geisselbasis der Spermatiden von *Marchantia polymorpha, Preissia quadrata, Sphaerocarpus Donnelli, Pellia Fabroniana* (Hepaticae). *Z. Naturf.*, **14b**, 399–401.

HODGE, A. J. (1959) Fine structure of lamellar systems as illustrated by chloroplasts. *In* ONCLEY, J. L. (ed.), *Biophysical Science — A Study Program,* pp. 331–341. John Wiley and Sons, New York.

——— (1960) Principles of ordering in fibrous systems. *Verhandl. IV intern. Kong. Elektronenmikroskopie, Berlin,* 1958, **2,** 119–139.

HOFFMANN-BERLING, H. (1955) Geisselmodelle und Adenosintriphosphat. *Biochim. Biophys. Acta.,* **16,** 146–154.

——— (1958a) Der Mechanismus eines neuen Kontraktionszyklus. *Biochim. Biophys. Acta.,* **27,** 247–255.

——— (1958b) The role of cell structures in cell movements. *Symp. Soc. Study Develop. Growth.,* **17,** 45–62.

HOLLANDE, A. (1952a) Classe des Chrysomonadines. *In* GRASSÉ, P. P. (ed.), *Traité de Zoologie,* T. 1, f., **1,** 471–570. Masson et Cie., Paris.

——— (1952b) Ordre des choanoflagellés ou craspedomonadines. *In* GRASSÉ, P. P. (ed.), *Traité de Zoologie,* T. 1, f., **1,** 579–598. Masson et Cie., Paris.

——— (1952c) Ordre des bodonides. *In* GRASSÉ, P. P. (ed.), *Traité de Zoologie,* T. 1, f., **1,** 669–693. Masson et Cie., Paris.

HOLTER, H. (1959) Pinocytosis. *Int. Rev. Cytol.,* **8,** 481–504.

HORNE, R. W. and B. A. NEWTON (1958) Intracellular structures in *Strigomonas oncopelti.* II. Fine structure of the kinetoplast-blepharoplast complex. *Exp. Cell Res.,* **15,** 103–111.

HOUWINK. A. L. (1951) An E. M. study of the flagellum of *Euglena gracilis. Proc. Acad. Sci. Amst.,* Ser. C, **54,** 132–137.

——— (1952) Die Pellikularschuppen und die Geissel der *Physomonas vestita* Stokes. *Z. wiss. Mikroskop.,* **60,** 402–404.

HOVASSE, R. and L. JOYON (1957) Sur l'ultrastructure de la Chrysomonadine *Hydrurus foetidus* Kirchner. *C. R. Acad. Sci., Paris,* **245,** 110–113.

HUGER, A. (1960) Electron microscope study on the cytology of a microsporidian spore by means of ultrathin sectioning. *J. Insect Path.,* **2,** 84–105.

HUTNER, S. H. (1961) The environment and growth: protozoan origins of metazoan responsivities. *Symp. Soc. gen. Microbiol.,* **10,** 1–18.

HUTNER, S. H. and L. PROVASOLI (1951) The phytoflagellates. *In* LWOFF, A. (ed.), *Biochemistry and Physiology of Protozoa,* **1,** 27–128. Academic Press, New York.

HYMAN, L. H. (1940) *The Invertebrates.* Vol. I. *Protozoa through Ctenophora.* McGraw-Hill, New York, 726 pp.

——— (1959) *The Invertebrates.* Vol. V. *Smaller Coelomate Groups.* McGraw-Hill, New York. 783 pp.

INABA, F. (1959) Electron-microscopic study on the fine structure of *Stentor coeruleus. Bull. biol. Soc. Hiroshima Univ.,* **10,** 35–43. (In Japanese with English summary.)

——— (1960) The fine structure of the nuclei of *Spirostomum ambiguum* seen by electron microscope. *Biol. J. Nara Women's Univ.,* **10,** 26–29.

INABA, F., NAKAMURA, R., and S. YAMAGUCHI (1958) An electron-microscopic study on the pigment granules of *Blepharisma. Cytologia, Tokyo,* **23,** 72–79.

INOKI, S., NAKANISHI, K., and T. NAKABAYASHI (1958) Study of *Leishmania donovani* with special reference to the kinetoplast, mitochondria, and Golgi zone by electron microscope employing the thin section technique. *Biken's J.*, **1**, 194–197.

———— (1959) Observations on *Trichomonas vaginalis* by electron microscopy. *Biken's J.*, **2**, 21–24.

INOKI, S., NAKANISHI, K., NAKABAYASHI, T., and M. OHNO (1957) Fine structure of the flagellum of *Leishmania donovani* revealed by electron-microscope. *Med. J. Osaka Univ.*, **7**, 719–729.

INOKI, S., OHNO, M., KONDO, K., and H. SAKAMOTO (1961) Electron-microscopic observations on the "costa" as one of the organelles in *Trichomonas foetus. Biken's J.*, **4**, 63–65.

INOUÉ, S. (1959) Motility of cilia and the mechanism of mitosis. *In* ONCLEY, J. L. (ed.), *Biophysical Science — A Study Program,* pp. 402–408. John Wiley and Sons, New York.

JAHN, B. (1953) Elektronenmikroskopische Untersuchungen an Foramini-ferenschalen. *Z. wiss. Mikroskop.*, **61**, 294–297.

JAHN, T. L. (1960) Ciliary reversal and activation by electric current; the Ludloff phenomenon in terms of core and volume conductors. *J. Protozool.*, **7** (Suppl.): 16.

JAHN, T. L., BOVEE, E. C., and E. B. SMALL (1960) The basis for a new major dichotomy of the Sarcodina. *J. Protozool.*, **7** (Suppl.): 8.

JAHN, T. L. and R. A. RINALDI (1959) Protoplasmic movement in the foraminiferan *Allogromia laticollaris* and a theory of its mechanism. *Biol. Bull. Wood's Hole,* **117**, 100–118.

JAKUS, M. A. (1945) The structure and properties of the trichocysts of *Paramecium. J. exp. Zool.*, **100**, 457–486.

———— (1956) Studies on the cornea. II. The fine structure of Descemet's membrane. *J. biophys. biochem. Cytol.*, **2** (Suppl.), 243–252.

JAKUS, M. A. and C. E. HALL (1946) Electron microscope observations of the trichocysts and cilia of *Paramecium. Biol. Bull., Wood's Hole,* **91**, 141–144.

JAROSCH, R. (1959) Zur Gleitbewegung der niederen Organismen. *Proto-plasma,* **50**, 277–289.

JONES, R. F. and R. A. LEWIN (1960) The chemical nature of the flagella of *Chlamydomonas moewusii. Exp. Cell Res.*, **19**, 408–410.

JURAND, A. (1961) An electron microscope study of food vacuoles in *Paramecium aurelia. J. Protozool.*, **8**, 125–130.

KASSEL, R. (1959) Particulates of amoebae. *Ann. N.Y. Acad. Sci.*, **78**, 421–431.

KERKUT, G. A. (1961) *Implications of Evolution.* Pergamon Press, London, 168 pp.

KILIAN, F. F. (1954) Die Feinstruktur des Kragens bei des Choanocyten der Spongilliden. *Bericht. Oberhess. Ges. Natur. Heilk. Giessen., Naturw. Abt.*, **27**, 85–89.

R

248 ELECTRON-MICROSCOPIC STRUCTURE OF PROTOZOA

KING, R. L., BEAMS, H. W., TAHMISIAN, T. N., and R. L. DEVINE (1961) The ciliature and infraciliature of *Nyctotherus ovalis* Leidy. *J. Protozool.,* **8,** 98–111.

KIRBY, H. (1944) Some observations on cytology and morphogenesis in flagellate protozoa. *J. Morphol.,* **75,** 361–421.

——— (1949) Systematic differentiation and evolution of flagellates in termites. *Rev. Soc. Mex. Hist. nat.,* **10,** 57–79.

KITCHING, J. A. (1954) On suction in Suctoria. *Colston Papers* (Proc. 7th Symp. Colston Res. Soc.), **7,** 197–203.

——— (1956a) Contractile vacuoles of protozoa. *Protoplasmatologia* III, D, **3a,** 1–45.

——— (1956b) Food vacuoles of protozoa. *Protoplasmatologia,* III, D, **3b,** 1–54.

KLEIN, B. M. (1942) Das Silverlinien oder neuroformative System der Ciliaten. *Ann. naturh.* (*Mus.*) *Hofmus., Wien,* **53,** 156–336.

KLEIN, R. L. and R. J. NEFF (1960) Osmotic properties of mitochondria isolated from *Acanthamoeba* sp. *Exp. Cell Res.,* **19,** 133–155.

KLEINSCHMIDT, A. and E. KINDER (1950) Elektronen-optische Befunde an Ratten-trypanosomen. *Zbl. Bakt.* Abt. I., **156,** 219–224.

KLUG, H. (1959) Cytomorphologische Untersuchungen an Gregarinen. *Biol. Zbl.,* **78,** 630–650.

KNOCH, M. and H. KÖNIG (1951) Zur Struktur der Paramecien-Trichocysten. *Naturwissenschaften,* **38,** 531.

KRAMER, J. P. (1960) Observations on the emergence of the microsporidian sporoplasm. *J. Insect Pathol.,* **2,** 433–439.

KRANEVELD, F. C., HOUWINK, A. L., and H. J. W. KEIDEL (1951) Electron microscopical investigations on trypanosomes. I. Some preliminary data regarding the structure of *Trypanosoma evansi. Proc. Acad. Sci. Amst.,* **54,** 393–399.

KRÜGER, F. and K. E. WOHLFARTH-BOTTERMANN (1952) Elektronoptische Beobachtungen an Ciliatenorganellen. *Mikroskopie,* **7,** 121–130.

KRÜGER, F., WOHLFARTH-BOTTERMANN, K. E., and G. PFEFFERKORN (1952) Protistenstudien III. Der Trichocysten von *Uronema marinum* Dujardin. *Z. Naturforsch.,* **7b,** 407–410.

KUDO, R. R. (1954) *Protozoology,* 4th ed. Charles C. Thomas, Springfield, Illinois. 966 pp.

——— (1959) *Pelomyxa* and related organisms. *Ann. N.Y. Acad. Sci.,* **78,** 474–486.

KUFF, E. L. and A. V. DALTON (1959) Biochemical studies of isolated Golgi membranes. *In* HAYASHI, T. (ed.), *Subcellular Particles,* pp. 114–127. Ronald Press, New York.

KÜMMEL. G. (1958) Die Gleitbewegung der Gregarinen. *Arch. Protistenk.,* **102,** 501–522.

LACKEY, J. B. (1940) Some new flagellates from the Woods Hole area. *Amer. Midl. Nat.,* **23,** 463–471.

LACY, D. D. and H. B. MILES (1959) Observations by electron microscopy on the structure of an acephaline gregarine (*Apolocystis elongata* Phillips and Mackinnon). *Nature, Lond.*, **183**, 1456–1457.

LAINSON, R., BAKER, J. R., BIRD, R. G., GARNHAM, P. C. C., and P. HEALEY (1961) Electron micrographs of *Lankesterella* (= *Atoxoplasma*) from English sparrows. *Trans. R. Soc. trop. Med. Hyg.*, **55**, 9.

LANDAU, J. V. (1959) Sol-gel transformations in amoebae. *Ann. N.Y. Acad. Sci.*, **78**, 487–500.

LANSING, A. I. and F. LAMY (1961) Fine structure of the cilia of rotifers. *J. biophys. biochem. Cytol.*, **9**, 799–812.

LASANSKY, A. and E. DEROBERTIS (1960) Submicroscopic analysis of the genetic distrophy of visual cells in C3H mice. *J. biophys. biochem. Cytol.*, **7**, 679–684.

LEHMANN, F. E. (1950) Globuläre Partikel als submikroskopische Elemente des tierischen Zytoplasmas. *Experientia*, **6**, 382–384.

—— (1958a) Der Feinbau der Organoide von *Amoeba proteus* und seine Beeinflussung durch verschiedene Fixierstoffe. *Ergebn. Biol.*, **21**, 88–127.

—— (1958b) Functional aspects of submicroscopic nuclear structures in *Amoeba proteus*, and of the mitotic apparatus of *Tubifex* embryos. *Exp. Cell Res.*, Suppl. **6**, 1–16.

LEHMANN, F. E., MANNI, E., and A. BAIRATI (1956) Der Feinbau von Plasmalemma und kontraktiler Vakuole bei *Amoeba proteus* in Schnitt- und Fragmentpräparaten. *Rev. suisse Zool.*, **63**, 246–255.

LEHMANN, F. E., MANNI, E., and W. GEIGER (1956) Der Schichtenbau des Plasmalemmas von *Amoeba proteus* im elektronenmikroskopischen Schnittbild. *Naturwissenschaften*, **43**, 91–92.

LEHNINGER, A. L. (1959) Relation of oxidation, phosphorylation and active transport to the structure of mitochondria. *In* ZIRKLE, R. (ed.), *A Symposium on Molecular Biology*, pp. 122–136. Univ. Chicago Press, Chicago.

LEVINE, L. (1960) Cytochemical adenosine triphosphatase of vorticellid myonemes. *Science*, **131**, 1377–1378.

LEWIN, R. A. (1958) The cell wall of *Platymonas*. *J. gen Microbiol.*, **19**, 87–90.

LÖFGREN, R. (1950) The structure of *Leishmania tropica* as revealed by phase and electron microscopy. *J. Bact.*, **60**, 617–625.

LOWNDES, A. G. (1944) The swimming of *Monas stigmatica* Pringsheim and *Peranema trichophorum* (Ehrbg.) Stein and *Volvox* sp. Additional experiments on the working of a flagellum. *Proc. zool. Soc., London*, III, **114**, 325–338.

LUDVIK, J. (1954) The study of the cell-morphology of *Trichomonas foetus* (Riedmüller) with the electron microscope. *Mém. Soc. zool. tchécosl.*, **18**, 189–197.

—— (1956) Vergleichende elektronoptische Untersuchungen an *Toxoplasma gondii* und *Sarcocystis tenella*. *Zbl. Bakt.* Abt. I, **166**, 60–65.

—— (1958) Elektronoptische Befunde zur Morphologie der Sarcosporidien (*Sarcocystis tenella* Railliet 1886). *Zbl. Bakt.* Abt. I, **172**, 330–350.

LUDVIK, J. (1960) The electron microscopy of *Sarcocystis miescheriana* Kuhn 1865. *J. Protozool.*, **7**, 128–135.

LWOFF, A. (1950) *Problems of Morphogenesis in Ciliates.* John Wiley and Sons, New York. 103 pp.

MANTON, I. (1952) The fine structure of plant cilia. *Symp. Soc. exp. Biol.*, **6**, 306–319.

—— (1953) Number of fibrils in the cilia of green algae. *Nature, Lond.*, **171**, 485–486.

—— (1955a) Observations with the electron microscope on *Synura caroliniana* Whitford. *Proc. Leeds philos. Soc.*, **6**, 306–316.

—— (1955b) Plane of bilateral symmetry in plant cilia. *Nature, Lond.*, **176**, 123.

—— (1956) Plant cilia and associated organelles. In RUDNICK, D. (ed.), *Cellular Mechanisms in Differentiation and Growth*, pp. 61–71. Princeton Univ. Press.

—— (1957a) Observations with the electron microscope on the cell structure of the antheridium and spermatozoid of *Sphagnum*. *J. exp. Bot.*, **8**, 382–400.

—— (1957b) Observations with the electron microscope on the internal structure of the zoospore of a brown alga. *J. exp. Bot.*, **8**, 294–303.

—— (1959a) Electron microscopical observations on a very small flagellate: the problem of *Chromulina pusilla* Butcher. *J. mar. biol. Ass. U.K.*, **38**, 319–333.

—— (1959b) Observations on the internal structure of the spermatozoid of *Dictyota*. *J. exp. Bot.*, **10**, 448–461.

—— (1959c) Observations on the microanatomy of the spermatozoid of the bracken fern (*Pteridium aquilinum*). *J. biophys. biochem. Cytol.*, **6**, 413–418.

MANTON, I. and B. CLARKE (1950) Electron microscope observations on the spermatozoid of *Fucus*. *Nature, Lond.*, **166**, 973.

—— (1951a) An electron microscope study of the spermatozoid of *Fucus serratus*. *Ann. Botan. N.S.*, **15**, 461–467.

—— (1951b) Demonstration of compound cilia in a fern spermatozoid with the electron microscope. *J. exp. Bot.*, **2**, 125–128.

—— (1951c) Electron microscope observations on the zoospores of *Pylaiella* and *Laminaria*. *J. exp. Bot.*, **2**, 242–246.

—— (1952) An electron microscope study of the spermatozoid of *Sphagnum*. *J. exp. Bot.*, **3**, 265–275.

—— (1956) Observations with the electron microscope on the internal structure of the spermatozoid of *Fucus*. *J. exp. Bot.*, **7**, 416–432.

MANTON, I., CLARKE, B., and A. D. GREENWOOD (1951) Observations with the electron microscope on a species of *Saprolegnia*. *J. exp. Bot.*, **2**, 321–331.

—— (1953) Further observations with the electron microscope on spermatozoids in the brown algae. *J. exp. Bot.*, **4**, 319–329.

—— (1955) Observations with the electron microscope on biciliate and quadriciliate zoospores in green algae. *J. exp. Bot.*, **6**, 126–128.

MANTON, I., CLARKE, B., GREENWOOD, A. D., and E. A. FLINT (1952) Further observations on the structure of plant cilia, by a combination of visual and electron microscopy. *J. exp. Bot.*, **3**, 204–215.

MANTON, I. and G. F. LEEDALE (1961a) Further observations on the fine structure of *Chrysochromulina ericina* Parke and Manton. *J. mar. biol. Ass. U.K.*, **41**, 145–155.

—— (1961b) Further observations on the fine structure of *Chrysochromulina minor* and *C. kappa* with special reference to the pyrenoids. *J. mar. biol. Ass. U.K.*, **41**, 519–526.

—— (1961c) Observations on the fine structure of *Paraphysomonas vestita*, with special reference to the Golgi apparatus and the origin of scales. *Phycologia*, **1**, 37–57.

MANTON, I. and M. PARKE (1960) Further observations on small green flagellates with special reference to possible relatives of *Chromulina pusilla* Butcher. *J. mar. biol. Ass. U.K.*, **39**, 275–298.

MAZIA, D., HARRIS, P. J., and T. BIBRING (1960) The multiplicity of the mitotic centers and the time-course of their duplication and separation. *J. biophys. biochem. Cytol.*, **7**, 1–20.

MERCER, E. H. (1959) An electron microscopic study of *Amoeba proteus*. *Proc. roy. Soc. B.*, **150**, 216–232.

MERCER, E. H. and B. M. SHAFFER (1960) Electron microscopy of solitary and aggregated slime mould cells. *J. biophys. biochem. Cytol.*, **7**, 353–356.

METZ, C. B., PITELKA, D. R., and J. A. WESTFALL (1953) The fibrillar systems of ciliates as revealed by the electron microscope. I. *Paramecium. Biol. Bull., Wood's Hole*, **104**, 408–425.

METZ, C. B. and J. A. WESTFALL (1954) The fibrillar systems of ciliates as revealed by the electron microscope. II. *Tetrahymena. Biol. Bull., Wood's Hole*, **107**, 106–122.

MEYER, H. and I. DE ANDRADE MENDONCA (1955) Electron microscopic observations of *Toxoplasma* " Nicolle et Manceaux " grown in tissue cultures. *Parasitology*, **45**, 449–451.

MEYER, H. and M. DE OLIVEIRA MUSACCHIO (1960) Electron microscope study of the exoerythrocytic form of *Plasmodium gallinaceum* in thin sections of infected tissue cultures. *J. Protozool.*, **7**, 222–228.

MEYER, H., DE OLIVEIRA MUSACCHIO, M., and I. DE ANDRADE MENDONCA (1958) Electron microscopic study of *Trypanosoma cruzi* in thin sections of infected tissue cultures and blood-agar forms. *Parasitology*, **48**, 1–8.

MEYER, H. and K. R. PORTER (1954) A study of *Trypanosoma cruzi* with the electron microscope. *Parasitology*, **44**, 16–23.

MEYER, H. and L. T. QUEIROGA (1960) Submicroscopical aspects of *Schizotrypanum cruzi* in thin sections of tissue culture forms. *J. Protozool.*, **7**, 124–127.

MILLER, J. H., SWARTZWELDER, J. C., and J. E. DEAS (1961) An electron microscopic study of *Entamoeba histolytica*. *J. Parasit.*, **47**, 577–587.

MOSSEVITCH, T. N. and E. M. CHEISSIN (1961) Certain data on electron microscope study of the merozoites of the *Eimeria intestinalis* from rabbit intestine. (In Russian). *Tsitologia*, **3**, 34–39.

MÜHLETHALER, K. (1956) Electron microscopic study of the slime mold *Dictyostelium discoideum*. *Amer. J. Bot.*, **43**, 673–678.

MÜHLPFORDT, H., and D. PETERS (1951) Elektronenmikroskopische Untersuchungen an Flagellatengeisseln von *Haematococcus pluvialis* (Flotow). *Zool. Anz.*, **15** (Suppl.), 153–159.

MÜLLER, M. and P. RÖHLICH (1961) Studies on feeding and digestion in Protozoa. II. Food vacuole cycle in *Tetrahymena corlissi*. *Acta Morphol. Acad. Sci. Hung.*, **10**, 297–305.

NAITOH, Y. (1958) Direct current stimulation of *Opalina* with intracellular microelectrode. *Annot. zool. Jap.*, **31**, 59–73.

NANNEY, D. L. and M. A. RUDZINSKA (1960) Protozoa. *In* BRACHET, J. and A. E. MIRSKY (eds.), *The Cell*, **4**, 109–150. Academic Press, New York.

NEEDHAM, A. E. (1959) The origination of life. *Quart. Rev. Biol.*, **34**, 189–209.

NELSON, L. (1959) Cytochemical studies with the electron microscope. II. Succinic dehydrogenase in rat spermatozoa. *Exp. Cell Res.*, **16**, 403–410.

NEMETSCHEK, T., HOFMANN, U., and K. E. WOHLFARTH-BOTTERMANN (1953) Die Querstreifung der *Paramecium*-Trichocysten. *Z. Naturforsch.*, **8b**, 383–384.

NEWTON, B. A. (1960) Fine structure of the kinetoplast in a trypanosomid flagellate. *Verhandl. IV intern. Kong. Elektronenmikroskopie, Berlin, 1958*, **2**, 515–517.

NEWTON, B. A. and R. W. HORNE (1957) Intracellular structures in *Strigomonas oncopelti*. I. Cytoplasmic structures containing ribonucleoprotein. *Exp. Cell Res.*, **13**, 563–574.

NOIROT-TIMOTHÉE, C. (1957) L'ultrastructure de l'appareil de Golgi des infusoires Ophryoscolecidae. *C. R. Acad. Sci., Paris.*, **244**, 2847–2849.

—— (1958a) Étude au microscope électronique des fibres retrociliaires des Ophryoscolecidae: leur ultrastructure, leur insertion, leur rôle possible. *C. R. Acad. Sci., Paris*, **246**, 1286–1289.

—— (1958b) L'ultrastructure de la limite ectoplasme-endoplasme et des fibres formant le caryophore chez les ciliés du genre *Isotricha* Stein (holotriches trichostomes). *C. R. Acad. Sci., Paris*, **247**, 692–695.

—— (1958c) L'ultrastructure du blépharoplaste des infusoires ciliés. *C. R. Acad. Sci., Paris*, **246**, 2293–2295.

—— (1958d) Quelques particularités de l'ultrastructure d'*Opalina ranarum* (Protozoa, Flagellata). *C. R. Acad. Sci., Paris.*, **247**, 2445–2447.

—— (1959) Recherches sur l'ultrastructure d'*Opalina ranarum*. *Ann. Sci. nat.*, **12**, 265–281.

—— (1960) Étude d'une famille de ciliés: les "Ophryoscolecidae". Structures et ultrastructures. *Ann. Sci. nat.*, **12**, 527–718.

NOLAND, L. E. (1957) Protoplasmic streaming: a perennial puzzle. *J. Protozool.*, **4**, 1–6.

NOVIKOFF, A. B. (1961) Mitochondria (chondriosomes). *In* BRACHET, J. and A. E. MIRSKY (eds.), *The Cell*, **2**, 299–421. Academic Press, New York.

OKAJIMA, A. (1953) Studies on the metachronal wave in *Opalina*. I. Electrical stimulation with the micro-electrode. *Jap. J. Zool.*, **11**, 87–100.

OKAJIMA, A. (1954a) Studies on the metachronal wave in *Opalina*. II. The regulating mechanism of ciliary metachronism and of ciliary reversal. *Annot. zool. Jap.*, **27**, 40–45.

—— (1954b) Studies on the metachronal wave in *Opalina*. III. Time-change of effectiveness of chemical and electrical stimuli during adaptation in various media. *Annot. zool. Jap.*, **27**, 46–51.

OSADA, M. (1959) Electron-microscopic studies on protozoa. I. Fine structure of *Entamoeba histolytica. Keijo J. Med.*, **8**, 99–103.

PALADE, G. E. (1959) Functional changes in the structure of cell components. In Hayashi, T. (ed.), *Subcellular Particles*, pp. 64–83. Ronald Press, New York.

PALADE, G. E. and P. SIEKEVITZ (1956a) Liver microsomes. An integrated morphological and biochemical study. *J. biophys. biochem. Cytol.*, **2**, 171–200.

—— (1956b) Pancreatic microsomes. An integrated morphological and biochemical study. *J. biophys. biochem. Cytol.*, **2**, 671–690.

PALAY, S. L. (ed.) (1958) *Frontiers in Cytology*, Yale Univ. Press, New Haven. 529 pp.

PAPPAS, G. D. (1956a) Helical structures in the nucleus of *Amoeba proteus. J. biophys. biochem. Cytol.*, **2**, 221–222.

—— (1956b) The fine structure of the nuclear envelope of *Amoeba proteus. J. biophys. biochem. Cytol.*, **2** (Suppl.), 431–434.

—— (1959 Electron microscope studies on amoebae. *Ann. N.Y. Acad. Sci.*, **78**, 448–473.

PAPPAS, G. D. and P. W. BRANDT (1959) Mitochondria. I. Fine structure of the complex patterns in the mitochondria of *Pelomyxa carolinensis* Wilson (*Chaos chaos* L.). *J. biophys. biochem. Cytol.*, **6**, 85–90.

PÁRDUCZ, B. (1954) Reizphysiologische Untersuchungen an Ziliaten. II. Neuere Beiträge zum Bewegungs- und Koordinations-mechanismus der Ziliatur. *Acta Biol. Acad. Sci. Hung.*, **5**, 169–212.

—— (1958a) Das interziliäre Fasernsystem in seiner Beziehung zu gewissen Fibrillenkomplexen der Infusorien. *Acta Biol. Acad. Sci. Hung.*, **8**, 191–218.

—— (1958b) Reizphysiologische Untersuchungen an Ziliaten. VII. Das Problem der vorbestimmten Leitungsbahnen. *Acta Biol. Acad. Sci. Hung.*, **8**, 219–251.

PARKE, M. and I. ADAMS (1960) The motile (*Crystallolithus hyalinas* Gaarder and Markali) and non-motile phases in the life history of *Coccolithus pelagicus* (Wallich) Schiller. *J. mar. biol. Ass. U.K.*, **39**, 263–274.

PARKE, M., MANTON, I., and B. CLARKE (1955) Studies on marine flagellates. II. Three new species of *Chrysochromulina. J. mar. biol. Ass. U.K.*, **34**, 579–609.

—— (1956) Studies on marine flagellates. III. Three further species of *Chrysochromulina. J. mar. biol. Ass. U.K.*, **35**, 387–414.

—— (1958) Studies on marine flagellates. IV. Morphology and micro-anatomy of a new species of *Chrysochromulina. J. mar. biol. Ass. U.K.*, **37**, 209–228.

PARKE, M., MANTON, I. and B. CLARKE (1959) Studies on marine flagellates. V. Morphology and microanatomy of *Chrysochromulina strobilus* sp. nov. *J. mar. biol. Ass. U.K.,* **38**, 169–188.

PAUTARD, F. G. E. (1959a) Adventitious formation of bone salts by algal flagella, and the question of apatite formation in protozoa generally. *Proc. XVth intern. Congr. Zool., London,* 1958, 478–479.

——— (1959b) Hydroxyapatite as a developmental feature of *Spirostomum ambiguum. Biochim. Biophys. Acta.,* **35**, 33–46.

PEASE, D. C. (1947) The structure of trichocysts revealed by the electron microscope. *J. cell. comp. Physiol.,* **29**, 91–94.

PETERSEN, J. B. and J. B. HANSEN (1954) Electron microscope observations on *Codonosiga botrytis* (Ehr.) James-Clark. *Botan. Tidsskr.,* **51**, 281–291.

——— (1956) On the scales of some *Synura* species. I. *Biol. Medd., Kbh.,* **23** (2), 3–28.

——— (1958) On the scales of some *Synura* species. II. *Biol. Medd., Kbh.,* **23** (7), 1–13.

——— (1960a) Elektronenmikroskopische Untersuchungen von zwei Arten der Heliozoen-Gattung *Acanthocystis. Arch. Protistenk.,* **104**, 547–553.

——— (1960b) On some neuston organisms. II. *Bot. Tidsskr,* **56**, 197–234.

PICKEN, L. E. R. (1960) *The Organization of Cells and Other Organisms.* Clarendon Press, Oxford. 629 pp.

PIGON, A. (1947) On the pellicle of *Euglena viridis* Ehrbg. *Bull. intern. Acad. Polon. Sci. Lett.* (Cracovie) (ser. B, 2) Année 1946, 111–120.

PITELKA, D. R. (1945) Morphology and taxonomy of flagellates of the genus *Peranema* Dujardin. *J. Morph.,* **76**, 179–192

——— (1949) Observations on flagellum structure in Flagellata. *Univ. Calif. Publ. Zool.,* **53**, 377–430.

——— (1956) An electron microscope study of cortical structures of *Opalina obtrigonoidea. J. biophys. biochem. Cytol.,* **2**, 423–432.

——— (1961a) Fine structure of the silverline and fibrillar systems of three tetrahymenid ciliates. *J. Protozool.,* **8**, 75–89.

——— (1961b) Observations on the kinetoplast-mitochondrion and the cytostome of *Bodo. Exp. Cell Res.,* **25**, 87–93.

PITELKA, D. R. and C. N. SCHOOLEY (1955) Comparative morphology of some protistan flagella. *Univ. Calif. Publ. Zool.,* **61**, 79–128.

——— (1957) The pellicular fine structure of *Euglena gracilis. J. Protozool,* **4** (Suppl.), 10.

——— (1958) The fine structure of the flagellar apparatus in *Trichonympha. J. Morph.,* **102**, 199–246.

PLAUT, W. (1959) Tracer studies in amoebae. *Ann. N.Y. Acad. Sci.,* **78**, 623–630.

POCHMANN, A. (1953) Struktur, Wachstum und Teilung der Körperhülle bei den Eugleninen. *Planta,* **42**, 478–548.

——— (1957) Zur Frage der ontogenetischen Gestaltentwicklung der Eugleninenzelle. *Ber. dtsch. bot. Ges.,* **70**, 291–296.

PORTER, K. R. (1957) The submicroscopic morphology of protoplasm. *Harvey Lectures,* **51**, 175–227.

PORTER, K. R. (1961) The ground substance; observations from electron microscopy. *In* BRACHET, J. and A. E. MIRSKY (eds.), *The Cell*, **2**, 621–676. Academic Press, New York.

PORTER, K. R. and F. L. KALLMAN (1952) Significance of cell particulates as seen by electron microscopy. *Ann. N.Y. Acad. Sci.*, **54**, 882–891.

POTTAGE, R. H. (1959) Electron microscopy of the adults and migrants of the suctorian ciliate *Discophrya piriformis*. *Proc. XVth intern. Congr. Zool. London*, 1958, 472–473.

POTTS, B. P. (1955) Electron microscope observations on trichocysts. *Biochim. Biophys. Acta*, **16**, 464–470.

POTTS, B. P. and S. G. TOMLIN (1955) The structure of cilia. *Biochim. Biophys. Acta*, **16**, 66–74.

PUYTORAC, P. DE (1952) Mise en évidence par le microscope électronique de plaques de revêtement à la surface du corps des Astomes. *C. R. Acad. Sci., Paris*, **234**, 2646–2648.

—— (1958) Origine infraciliaire des fibres squélettiques de certains infusoires astomes et présence d'un ergastoplasme chez ces ciliés. *C. R. Acad. Sci., Paris*, **246**, 3186–3188.

—— (1959a) Le cytosquelette et les systèmes fibrillaires du cilié *Metaradiophrya gigas* de Puytorac, d'après étude au microscope électronique. *Arch. Anat. microscop.*, **48**, 49–62.

—— (1959b) Structures et ultrastructures nucléolaires et perinucléolaires du macronoyau intermitotique des ciliés Haptophryidae. *C. R. Acad. Sci., Paris*, **249**, 1709–1711.

—— (1960) Observations en microscope électronique de l'appareil vacuolaire pulsatile chez quelques ciliés astomes. *Arch Anat. microscop.*, **49**, 241–256.

—— (1961a) Complément a l'étude de l'ultrastructure des ciliés du genre *Metaradiophrya* Heid. 1935. *Arch. Anat. microscop.*, **50**, 35–58.

—— (1961b) Nouvelles observations sur l'argyrome des ciliés astomes par l'emploi du microscope électronique. *C. R. Ass. Anat.*, **109**, 675–678.

PYNE, C. K. (1958) Electron microscopic investigations on the leptomonad form of *Leishmania donovani*. *Exp. Cell Res.*, **14**, 388–397.

—— (1959) L'ultrastructure de *Cryptobia helicis* (Flagellé, Fam. Bodonidae). *C. R. Acad. Sci., Paris*, **248**, 1410–1413.

—— (1960a) Études sur la structure inframicroscopique du cinétoplaste chez *Leishmania tropica*. *C. R. Acad. Sci., Paris*, **251**, 2776–2778.

—— (1960b) L'ultrastructure de l'appareil basal des flagelles chez *Cryptobia helicis* (Flagellé, Bodonidae). *C. R. Acad. Sci., Paris.*, **250**, 1912.

PYNE, C. K. and J. CHAKRABORTY (1958) Electron microscopic studies on the basal apparatus of the flagellum in the protozoon, *Leishmania donovani*. *J. Protozool.*, **5**, 264–268.

RANDALL, J. T. (1956) Fine structure of some ciliate protozoa. *Nature, Lond.*, **178**, 9–14.

—— (1957) The fine structure of the protozoan *Spirostomum ambiguum*. *Symp. Soc. exp. Biol.*, **10**, 185–198.

RANDALL, J. T. (1959a) Contractility in the stalks of Vorticellidae. *J. Protozool.*, 6 (Suppl.), 30.

—— (1959b) The nature and significance of kinetosomes. *J. Protozool.*, 6 (Suppl.), 30.

—— (1959c) The stalks of Epistylidae. *J. Protozool.*, 6 (Suppl.), 30–31.

RANDALL, J. T. and S. F. JACKSON (1958) Fine structure and function in *Stentor polymorphus. J. biophys. biochem. Cytol.*, 4, 807–830.

RASMONT, R. (1959) L'ultrastructure des choanocytes d'éponges. *Ann. Sci. Nat. Zool.*, 12, 253–262.

RAY, H. N., DAS GUPTA, N. N., DE, M. L., and A. GUHA (1955) A new structure observed in *Trypanosoma evansi* (Indian strain). *Nature, Lond.*, 175, 392–393.

REGER, J. F. and H. W. BEAMS (1954) Electron micrographs of the pellicle of a species of *Euglena. Proc. Iowa Acad. Sci.*, 61, 593–597.

RHODIN, J. and T. DALHAMN (1956) Electron microscopy of the tracheal ciliated mucosa in rat. *Z. Zellforsch.*, 44, 345–412.

RIS, H. and R. N. SINGH (1961) Electron microscope studies on blue-green algae. *J. biophys. biochem. Cytol.*, 9, 63–80.

ROBERTSON, J. D. (1960) A molecular theory of cell membrane structure. *Verhandl. IV intern. Kong. Elektronenmikroskopie, Berlin*, 1958, 2, 159–171.

ROBINOW, C. F. (1956) Observations on vase-shaped, iron-containing houses of two colorless flagellates of the family Bicoecidae. *J. biophys. biochem. Cytol.*, 4 (Suppl.), 233–234.

—— (1960) Outline of the visible organization of bacteria. *In* BRACHET, J. and A. E. MIRSKY (eds.), *The Cell*, 4, 45–108. Academic Press, New York.

ROTH, L. E. (1956) Aspects of ciliary fine structure in *Euplotes patella. J. biophys. biochem. Cytol.*, 2 (Suppl.), 235–242.

—— (1957) An electron microscope study of the cytology of the protozoan *Euplotes patella. J. biophys. biochem. Cytol.*, 3, 985–1000.

—— (1958a) A filamentous component of protozoan fibrillar systems. *J. Ultrastruct. Res.*, 1, 223–234.

—— (1958b) Ciliary coordination in the protozoa. *Exp. Cell Res.*, Suppl. 5, 573–585.

—— (1959) An electron-microscope study of the cytology of the protozoan *Peranema trichophorum. J. Protozool.*, 6, 107–116.

—— (1960a) Electron microscopy of pinocytosis and food vacuoles in *Pelomyxa. J. Protozool.*, 7, 176–185.

—— (1960b) Observations on division stages in the protozoan hypotrich *Stylonychia. Verhandl. IV intern. Kong. Elektronenmikroskopie, Berlin*, 1958, 2, 241–244.

ROTH, L. E. and E. W. DANIELS (1961) Infective organisms in the cytoplasm of *Amoeba proteus. J. biophys. biochem. Cytol.*, 9, 317–323.

ROTH, L. E. and O. T. MINICK (1961) Electron microscopy of nuclear and cytoplasmic events during division in *Tetrahymena pyriformis* strains W and HAM 3. *J. Protozool.*, 8, 12–21.

ROTH, L. E., OBETZ, S. W., and E. W. DANIELS (1960) Electron microscopic studies of mitosis in amebae. I. *Amoeba proteus*. *J. biophys. biochem. Cytol.*, **8**, 207–220.

ROUILLER, C. (1960) Physiological and pathological changes in mitochondrial morphology. *Int. Rev. Cytol.*, **9**, 227–292.

ROUILLER, C. and E. FAURÉ-FREMIET (1957a) L'ultrastructure des trichocystes fusiformes de *Frontonia atra*. *Bull. Microscop. appl.*, **7**, 135–139.

—— (1957b) Ultrastructure réticulée d'une fibre squelettique chez un cilié. *J. Ultrastruct. Res.*, **1**, 1–13.

—— (1958a) Structure fine d'un flagellé chrysomonadien: *Chromulina psammobia*. *Exp. Cell Res.*, **14**, 47–67.

—— (1958b) Ultrastructure des cinétosomes à l'état de repos et à l'état cilifère chez un cilié péritriche. *J. Ultrastruct. Res.*, **1**, 289–294.

ROUILLER, C., FAURÉ-FREMIET, E., and M. GAUCHERY (1956a) Fibres scléroprotéiques d'origine ciliaire chez les infusoires péritriches. *C. R. Acad. Sci., Paris*, **242**, 180–182.

—— (1956b) Les tentacules d'*Ephelota;* étude au microscope électronique. *J. Protozool.*, **3**, 194–200.

—— (1956c) Origine ciliaire des fibrilles scléroprotéiques pédonculaires chez les ciliés péritriches. *Exp. Cell Res.*, **11**, 527–541.

—— (1956d) The pharyngeal protein fibers of the ciliates. *Proc. Stockholm Conf. Electron Microscopy*, pp. 216–218.

RUDZINSKA, M. A. (1956) Further observations on the fine structure of the macronucleus in *Tokophrya infusionum*. *J. biophys. biochem. Cytol.*, **2**, 425–430.

—— (1958) An electron microscope study of the contractile vacuole in *Tokophrya infusionum*. *J. biophys. biochem. Cytol.*, **4**, 195–202.

RUDZINSKA, M. A., BRAY, R. S., and W. TRAGER (1960) Intracellular phagotrophy in *Plasmodium falciparum* and *Plasmodium gonderi*. *J. Protozool.*, **7**, (Suppl.), 24–25.

RUDZINSKA, M. A. and K. R. PORTER (1954a) Electron microscope study of intact tentacles and disc in *Tokophrya infusionum*. *Experientia*, **10**, 460–467.

—— (1954b) The fine structure of *Tokophrya infusionum* with emphasis on the feeding mechanism. *Trans. N.Y. Acad. Sci.*, Ser. 2, **16**, 408–411.

—— (1955) Observations on the fine structure of the macronucleus of *Tokophrya infusionum*. *J. biophys. biochem. Cytol.*, **1**, 421–428.

RUDZINSKA, M. A. and W. TRAGER (1957) Intracellular phagotrophy by malaria parasites: an electron microscope study of *Plasmodium lophurae*. *J. Protozool.*, **4**, 190–199.

—— (1959) Phagotrophy and two new structures in the malaria parasite *Plasmodium berghei*. *J. biophys. biochem. Cytol.*, **6**, 103–112.

—— (1960) The fine structure of *Babesia rodhaini*. *J. Protozool.* **7** (Suppl.), 11.

—— (1961) The role of the cytoplasm during reproduction in a malarial parasite (*Plasmodium lophurae*) as revealed by electron microscopy. *J. Protozool.*, **8**, 307–322.

SAGER, R. (1959) The architecture of the chloroplast in relation to its photosynthetic activities. *Brookhaven Symp. Biol.*, **11**, 101–116.

SAGER, R. and G. E. PALADE (1954) Chloroplast structure in green and yellow strains of *Chlamydomonas*. *Exp. Cell Res.*, **7**, 584–588.

—— (1957) Structure and development of the chloroplast in *Chlamydomonas*. I. The normal green cell. *J. biophys. biochem. Cytol.*, **3**, 463–488.

SCHAEFFER, A. A. (1926) Taxonomy of the amebas. *Papers Dept. Marine Biol. Carnegie Inst. Wash.*, **24**, 1–116.

SCHMITT, F. O. (1960) Electron microscopy in morphology and molecular biology. *Verhandl. IV intern. Kong. Elektronenmikroskopie, Berlin, 1958,* **2**, 1–16.

SCHMITT, F. O., HALL, C. E., and M. A. JAKUS (1943) The ultrastructure of protoplasmic fibrils. *Biol. Symp.*, **10**, 261–276.

SCHNEIDER, L. (1959) Neue Befunde über den Feinbau des Cytoplasmas von *Paramaecium* nach Einbettung in Vestopal W. *Z. Zellforsch*, **50**, 61–77.

—— (1960a) Die Feinstruktur des Nephridialplasmas von *Paramaecium*. *Zool. Anz.*, **23** (Suppl.), 457–470.

—— (1960b) Elektronenmikroskopische Untersuchungen über das Nephridialsystem von *Paramaecium*. *J. Protozool.*, **7**, 75–90.

SCHNEIDER, L. and K. E. WOHLFARTH-BOTTERMANN (1959) Protistenstudien IX. Elektronenmikroskopische Untersuchungen an Amöben unter besonderer Berücksichtigung der Feinstruktur des Cytoplasmas. *Protoplasma*, **51**, 377–389.

SCHOLTYSECK, E. and N. WEISSENFELS (1956) Elektronenmikroskopische Untersuchungen von Sporozoen I. Die Oocystenmembran des Hühnercoccids *Eimeria tenella*. *Arch. Protistenk.*, **101**, 215–222.

SCHUMAKER, V. N. (1958) Uptake of protein from solution by *Amoeba proteus*. *Exp. Cell Res.*, **15**, 314–331.

SCHUSSNIG, B. (1960) *Handbuch der Protophytenkunde*. Bd. II, Gustav Fischer, Jena. 1144 pp.

Seaman, G. R. (1960) Large-scale isolation of kinetosomes from the ciliated protozoan *Tetrahymena pyriformis*. *Exp. Cell Res.*, **21**, 292–302.

SEDAR, A. W. and K. R. PORTER (1955) The fine structure of cortical components of *Paramecium multimicronucleatum*. *J. biophys. biochem. Cytol.*, **1**, 583–604.

SEDAR, A. W. and M. A. RUDZINSKA (1956) Mitochondria of protozoa. *J. biophys. biochem. Cytol.*, **2** (Suppl.), 331–336.

SERRA, J. A. (1960) Why flagella and cilia have $1+9$ pairs of fibers. *Exp. Cell Res.*, **20**, 395–400.

SJÖSTRAND, F. S. (1959) Fine structure of cytoplasm: the organization of membranous layers. *In* Oncley, J. L. (ed.), *Biophysical Science — A Study Program*, pp. 301–318. John Wiley and Sons, New York.

SLAUTTERBACK, D. B. and D. W. FAWCETT (1959) The development of the cnidoblasts of *Hydra*. *J. biophys. biochem. Cytol.*, **5**, 441–452.

SLEIGH, M. A. (1956) Metachronism and frequency of beat in the peristomial cilia of *Stentor*. *J. exp. Biol.*, **33**, 15–28.

SLEIGH, M. A. (1957) Further observations on coordination and the determination of frequency in the peristomial cilia of *Stentor*. *J. exp. Biol.*, **34,** 106–115.

SMITH, G. M. (1955) *The Fresh-water Algae of the United States*. 2nd ed. McGraw-Hill, New York, 716 pp.

SONNEBORN, T. M. (1949) Ciliated protozoa: cytogenetics, genetics and evolution. *Annu. Rev. Microbiol.*, **3,** 55–96.

SOTELO, J. R. and O. TRUJILLO-CENÓZ (1958a) Electron microscope study on the development of ciliary components of the neural epithelium of the chick embryo. *Z. Zellforsch.*, **49,** 1–12.

—— (1958b) Electron microscope study of the kinetic apparatus in animal sperm cells. *Z. Zellforsch.*, **48,** 565–601.

—— (1959) The fine structure of an elementary contractile system. *J. biophys. biochem. Cytol.*, **6,** 126–128.

STEINERT, G., FIRKET, H. and M. STEINERT (1958) Synthèse d'acide désoxyribonucléique dans le corps parabasal de *Trypanosoma mega*. *Exp. Cell Res.*, **15,** 632–634.

STEINERT, M. (1960) Mitochondria associated with the kinetonucleus of *Trypanosoma mega*. *J. biophys. biochem. Cytol.*, **8,** 542–546.

STEINERT, M. and A. B. NOVIKOFF (1960) The existence of a cytostome and the occurrence of pinocytosis in the trypanosome (*Trypanosoma mega*). *J. biophys. biochem. Cytol.*, **8,** 563–570.

STEINHAUS, E. A. (1951) Report on diagnoses of diseased insects 1944–1950. *Hilgardia*, **20,** 629–678.

STENZEL, H. B. (1950) Proposed uniform endings for names of higher categories in zoological systematics. *Science,* **112,** 94.

STEWART, P. A. and B. T. STEWART (1959) Protoplasmic streaming and the fine structure of slime mold plasmodia. *Exp. Cell Res.*, **18,** 374–377.

—— (1961) Membrane formation during sclerotization of *Physarum polycephalum* plasmodia. *Exp. Cell Res.*, **23,** 471–478.

SZENT-GYÖRGYI, A. (1960) *Introduction to a Submolecular Biology*. Academic Press, New York. 135 pp.

TARTAR, V. (1960) Reconstitution of minced *Stentor coeruleus*. *J. exp. Zool.*, **144,** 187–208.

—— (1961) *The Biology of Stentor*. Pergamon Press, London. 413 pp.

TAYLOR, C. V. (1941) Fibrillar systems in ciliates. *In* CALKINS, G. N. and F. M. SUMMERS (eds.), *Protozoa in Biological Research*, pp. 191–270. Columbia Univ. Press, New York.

THÉODORIDÈS, J. (1959) Étude des eugrégarines au microscope électronique. *Proc. XVth intern. Congr. Zool. London,* 1958, 477–478.

TIBBS, J. (1957) The nature of algal and related flagella. *Biochim. Biophys. Acta*, **23,** 275–288.

—— (1958) The properties of algal and sperm flagella obtained by sedimentation. *Biochim. Biophys. Acta*, **28,** 636–637.

TOKUYASU, K. and E. YAMADA (1959) The fine structure of the retina studied with the electron microscope. IV. Morphogenesis of outer segments of retinal rods. *J. biophys. biochem. Cytol.*, **6,** 225–230.

TRAGER, W. (1960) Intracellular parasitism and symbiosis. In BRACHET, J. and A. E. MIRSKY (eds.), The Cell, 4, 151–214. Academic Press, New York.

Ts'o, P. O. P., Jr., BONNER, J., EGGMAN, L. and J. VINOGRAD (1956) Observations on an ATP-sensitive protein system from the plasmodia of a myxomycete. J. gen. Physiol., 39, 325–347.

TSUJITA, M., WATANABE, K., and S. TSUDA (1954) Electron microscopical studies on the inner structure of Paramecium caudatum by means of ultra-thin sections. Cytologia, Tokyo, 19, 306–316.

—— (1957) Electron microscopy of thin-sectioned nuclei in Paramecium. Cytologia, Tokyo, 22, 322–327.

TUFFRAU, M. (1960) Révision du genre Euplotes, fondée sur la comparaison des structures superficielles. Hydrobiologia, 15, 1–77.

UEDA, K. (1958) Structure of plant cells with special reference to lower plants. III. A cytological study of Euglena gracilis. Cytologia, Tokyo, 23, 56–67.

VICKERMAN, K. (1960) Structural changes in mitochondria of Acanthamoeba at encystation. Nature, Lond., 188, 248–249.

VILLENEUVE-BRACHON, S. (1940) Recherches sur les ciliés hétérotriches. Arch. Zool. exptl. gén., 82, 1–180.

VIVIER, E. and J. ANDRÉ (1961) Existence d'inclusions d'ultrastructure fibrillaire dans le macronucleus de certaines souches de Paramecium caudatum. C. R. Acad. Sci., Paris, 252, 1848–1850.

VLK, W. (1938) Über den Bau der Geissel. Arch. Protistenk., 90, 448–488.

WATSON, M. R., HOPKINS, J. M., and J. T. RANDALL (1961) Isolated cilia from Tetrahymena pyriformis. Exp. Cell. Res., 23, 629–631.

WEISER, J. (1959) Nosema laphygmae n. sp. and the internal structure of the microsporidian spore. J. Insect Pathol., 1, 52–59.

WEISZ, P. B. (1954) Morphogenesis in protozoa. Quart. Rev. Biol., 29, 207–229.

—— (1956) Experiments on the initiation of division in Stentor coeruleus. J. exp. Zool., 131, 137–162.

WELLINGS, S. R. and K. B. DEOME (1961) Milk protein droplet formation in the Golgi apparatus of the C3H/Crgl mouse mammary epithelial cells. J. biophys. biochem. Cytol., 9, 479–485.

WELLINGS, S. R., DEOME, K. B., and D. R. PITELKA (1960) Electron microscopy of milk secretion in the mammary gland of the C3H/Crgl mouse. I. Cytomorphology of the prelactating and the lactating gland. J. Nat. Cancer Inst., 25, 393–421.

WESSENBERG, H. (1961) Studies on the life cycle and morphogenesis of Opalina. Univ. Calif. Publ. Zool., 61, 315–370.

WETTSTEIN, D. v. (1957) Developmental changes in chloroplasts and their genetic control. In RUDNICK, D. (ed.), Developmental Cytology, pp.123–160. Ronald Press, New York.

WICHTERMAN, R. (1953) The Biology of Paramecium. Blakiston, New York. 527 pp.

WOHLFARTH-BOTTERMANN, K. E. (1953a) Experimentelle und elektronenoptische Untersuchungen zur Funktion der Trichocysten von
Paramecium caudatum. Arch. Protistenk., **98**, 169–226.
—— (1953b) Protistenstudien IV. Weitere Untersuchungen an Ciliaten-
Cilien. *Zool. Anz.*, **17** (Suppl.), 335–346.
—— (1956) Protistenstudien VII. Die Feinstruktur der Mitochondrien
von *Paramecium caudatum. Z. Naturf.*, **11b**, 578–581.
—— (1958a) Cytologische Studien II. Die Feinstruktur des Cytoplasmas
von *Paramecium. Protoplasma*, **49**, 231–247.
—— (1958b) Cytologische Studien V. Feinstrukturveränderungen des
Cytoplasmas und der Mitochondrien von *Paramecium* nach Einwirkung
letaler Temperaturen und Röntgendosen. *Protoplasma*, **50**, 82–92.
—— (1958c) Elektronen-mikroskopie. Neue Erkenntnisse zum Feinbau
der Zelle. *Umschau*, **5**, 144–147.
—— (1959a) Gestattet das elektronenmikroskopische Bild Aussagen zur
Dynamik in der Zelle? Cytologische Studien VI. *Z. Zellforsch.*, **50**, 1–27.
—— (1959b) Protistenstudien VIII. Weitere Untersuchungen zur Feinstruktur der Axopodien von Heliozoen. *Zool. Anz.*, **163**, 1–10.
—— (1960a) Die elektronenmikroskopische Untersuchung cytoplasma-
tischer Strukturen. *Zool. Anz.*, **23** (Suppl.), 393–419.
—— (1960b) Protistenstudien X. Licht- und elektronenmikroskopische
Untersuchungen an der Amöbe *Hyalodiscus simplex* n. sp. *Protoplasma*,
52, 58–107.
—— (1961) Cytologische Studien VII. Strukturaspekte der Grundsub-
stanz des Cytoplasmas nach Einwirkung verschiedener Fixierungsmittel.
Protoplasma, **53**, 259–290.
WOHLFARTH-BOTTERMANN, K. E. and F. KRÜGER (1954) Protistenstudien
VI. Die Feinstruktur der Axopodien und der Skelettnadeln von
Heliozoen. *Protoplasma*, **43**, 177–191.
WOHLFARTH-BOTTERMANN, K. E. and A. PFEFFERKORN (1953) Protistenstudien V. Zur Struktur des Wimperapparates. *Protoplasma*, **42**,
227–238.
WOLKEN, J. J. (1956) A molecular morphology of *Euglena gracilis* var.
bacillaris. J. Protozool., **3**, 211–221.
—— (1959) The chloroplast structure, pigments, and pigment-protein
complex. *Brookhaven Symp. Biol.*, **11**, 87–99.
—— (1960) Photoreceptor structures. *Norelco Reporter*, **7**, 62–66.
WOLKEN, J. J. and G. E. PALADE (1953) An electron microscope study of
two flagellates. Chloroplast structure and variation. *Ann. N. Y. Acad.
Sci.*, **56**, 873–889.
YAGIU, R. and Y. SHIGENAKA (1958a) Electron microscopical observation
of *Condylostoma spatiosum* Osaki and Yagiu in ultra-thin section. I. The
pellicle and vacuoles in the cortex. *Dobutsugaku Zasshi*, **67**, 106–110.
—— (1958b) Electron microscopical observation of *Condylostoma spatiosum*
Osaki and Yagiu in ultra-thin section. II. The cilium and basal granule.
Dobutsugaku Zasshi, **67**, 158–162.

YAGIU, R. and Y. SHIGENAKA (1958c) Observations of ectoplasmic components of *Condylostoma spatiosum* with electron microscope. *J. Electronmicroscopy* (Chiba), **6**, 38–41.

—— (1959a) Electron microscopical observation of *Condylostoma spatiosum* Osaki and Yagiu in ultra-thin section. III. A relative position between the basal granules and the longitudinal fibrillar bundle. *Dobutsugaku Zasshi*, **68**, 8–13.

—— (1959b) Electron microscopical observation of *Condylostoma spatiosum* Osaki and Yagiu in ultra-thin section. IV. The fibrils between the basal granule and the longitudinal fibrillar bundle. *Dobutsugaku Zasshi*, **68**, 414–418.

YASUZUMI, G. and H. ISHIDA (1957) Spermatogenesis in animals as revealed by electron microscopy. II. Submicroscopic structure of developing spermatid nuclei of grasshopper. *J. biophys. biochem. Cytol.*, **3**, 663–668.

ZAMAN, V. (1961) An electron-microscopic observation of the "tail" end of *Entamoeba invadens*. *Trans. R. Soc. trop. Med. Hyg.*, **55**, 263–264.

INDEX

263